D1234825

The Politics and Pleasures of Consuming Differently

Consumption and Public Life

Series Editors: Frank Trentmann and Richard Wilk

Titles include:

Mark Bevir and Frank Trentmann (*editors*)
GOVERNANCE, CITIZENS AND CONSUMERS
Agency and Resistance in Contemporary Politics

Magnus Boström and Mikael Klintman
ECO-STANDARDS, PRODUCT LABELLING AND GREEN CONSUMERISM

Daniel Thomas Cook (*editor*)
LIVED EXPERIENCES OF PUBLIC CONSUMPTION
Encounters with Value *in* Marketplaces on Five Continents

Nick Couldry, Sonia Livingstone and Tim Markham
MEDIA CONSUMPTION AND PUBLIC ENGAGEMENT
Beyond the Presumption of Attention

Amy Randall
THE SOVIET DREAM WORLD OF RETAIL TRADE AND CONSUMPTION IN THE 1930s

Kate Soper, Martin Ryle and Lyn Thomas (*editors*)
THE POLITICS AND PLEASURES OF CONSUMING DIFFERENTLY

Kate Soper and Frank Trentman (*editors*)
CITIZENSHIP AND CONSUMPTION

Harold Wilhite
CONSUMPTION AND THE TRANSFORMATION OF EVERYDAY LIFE
A View from South India

Forthcoming:

Jacqueline Botterill
CONSUMER CULTURE AND PERSONAL FINANCE
Money Goes to Market

Roberta Sassatelli
FITNESS CULTURE
Gyms and the Commercialisation of Discipline and Fun

Consumption and Public Life
Series Standing Order ISBN 978–1–4039–9983–2 Hardback
978–1–4039–9984–9 Paperback
(*outside North America only*)

You can receive future titles in this series as they are published by placing a standing order. Please contact your bookseller or, in case of difficulty, write to us at the address below with your name and address, the title of the series and the ISBN quoted above.

Customer Services Department, Macmillan Distribution Ltd, Houndmills, Basingstoke, Hampshire RG21 6XS, England

---

The Politics and Pleasures of Consuming Differently

Edited By

Kate Soper
London Metropolitan University, UK

Martin Ryle
University of Sussex, UK

and

Lyn Thomas
London Metropolitan University, UK

First published 2009 by
PALGRAVE MACMILLAN

Palgrave Macmillan in the UK is an imprint of Macmillan Publishers Limited,
registered in England, company number 785998, of Houndmills, Basingstoke,
Hampshire RG21 6XS.

Palgrave Macmillan in the US is a division of St Martin's Press LLC,
175 Fifth Avenue, New York, NY 10010.

Palgrave Macmillan is the global academic imprint of the above companies
and has companies and representatives throughout the world.

Palgrave® and Macmillan® are registered trademarks in the United States,
the United Kingdom, Europe and other countries.

ISBN-13: 978–0–230–53728–6 hardback
ISBN-10: 0–230–53728–6 hardback

This book is printed on paper suitable for recycling and made from fully
managed and sustained forest sources. Logging, pulping and manufacturing
processes are expected to conform to the environmental regulations of the
country of origin.

A catalogue record for this book is available from the British Library.

Library of Congress Cataloging-in-Publication Data
The politics and pleasures of consuming differently / [edited by]
Kate Soper, Martin Ryle, Lyn Thomas.
p. cm. — (Consumption and public life)
Includes bibliographical references and index.
ISBN 978–0–230–53728–6
1. Consumption (Economics)—Social aspects.
2. Consumption (Economics)—Environmental aspects. 3. Consumption
(Economics)—Moral and ethical aspects. 4. Shopping. 5. Quality of
life. I. Soper, Kate. II. Ryle, Martin H. III. Thomas, Lyn, 1953–
HC79.C6P628 2009
339.4′7—dc22 2008031478

10 9 8 7 6 5 4 3 2 1
18 17 16 15 14 13 12 11 10 09

Printed and bound in Great Britain by
CPI Antony Rowe, Chippenham and Eastbourne

Contents

Notes on Contributors vii

Introduction: The Mainstreaming of Counter-Consumerist Concern 1
Kate Soper

Part I Texts and Representations

1 Representing Consumers: Contesting Claims and Agendas 25
Roberta Sassatelli

2 The Past, the Future and the Golden Age: Some Contemporary Versions of Pastoral 43
Martin Ryle

3 Ecochic: Green Echoes and Rural Retreats in Contemporary Lifestyle Magazines 59
Lyn Thomas

4 Mediated Culture and Exemptionalism 74
Simon Blanchard

Part II Value, Hedonism, Critique

5 The Bohemian Habitus: New Social Theory and Political Consumerism 93
Sam Binkley

6 Sustainable Hedonism: The Pleasures of Living within Environmental Limits 113
Marius de Geus

7 Green Pleasures 130
Richard Kerridge

Part III Everyday Consumption

8 Happiness and the Consumption of Mobility 157
Juliet Solomon

9 Gendering Anti-Consumerism: Alternative Genealogies, Consumer Whores and the Role of *Ressentiment* 171
Jo Littler

10 Growing Sustainable Consumption Communities: The Case of Local Organic Food Networks 188
Gill Seyfang

Part IV Conclusion

Conclusion 209
Martin Ryle, Kate Soper and Lyn Thomas

Index 213

Notes on Contributors

Sam Binkley is Assistant Professor of Sociology at Emerson College. His research focuses on questions of identity formation in the context of contemporary cultures of consumption, with a concentration on the role of lifestyle movements, from the counterculture of the 1960s and 1970s to contemporary anti-consumerist activism. His recent monograph, *Loosening Up: Lifestyle Consumption in the 1970s* (Duke University Press, 2007), examines the role of lifestyle discourse in the shaping of reflexive identity. He has also published on a range of theoretical and empirical topics, including the relevance of Michel Foucault to the study of contemporary consumerism, and the culture of consumption in contemporary Cuba. His articles have appeared in *The Journal of Consumer Culture; Cultural Studies – Critical Methodologies; Cultural Studies;* and *The Journal for Cultural Research.*

Simon Blanchard is Senior Lecturer in Democracy and Media in the School of Cultural Studies, Leeds Metropolitan University. Prior to that he was Senior Research Fellow with the Centre for British Film and Television Studies, and has been Lecturer and Research Fellow at Manchester Metropolitan and Sheffield Hallam Universities. From 1980 to 1988 he was National Organiser for the Independent Film, Video and Photography Association. Recent publications include a survey of media activist initiatives for *The Alternative Media Handbook* (Routledge, forthcoming) and a study of TV export policies included in Sylvia Harvey (ed.), *Trading Culture: Global Traffic and Local Cultures in Film and Television* (John Libbey, 2006).

Marius de Geus is Senior Lecturer in the Political Science Department at the University of Leiden, where he teaches political theory and environmental philosophy. He has published extensively on green politics and green political theory, particularly on the ecological restructuring of the state, ecological utopias and the environmental debate, landscape change in Western Europe, and on sustainability and liberalism. His books include *Politics, Environment and Liberty* (Van Arkel,1993), *Democracy and Green Political Thought* (edited with Brian Doherty, Routledge,

1996), *Ecological Utopias: Envisioning the Sustainable Society* (International Books,1999), *Fences and Freedom: The Philosophy of Hedgelaying* (with Thomas van Slobbe, International Books, 2003) and *The End of Over-consumption: Towards a lifestyle of Moderation and Self-restraint* (International Books, 2003).

Richard Kerridge is Head of Postgraduate Studies in English Literature and Creative Writing at Bath Spa University, where he teaches both subjects. He has published and lectured widely on contemporary literature and writing and environmentalism, and has played an important role in the development of eco-critical teaching and writing in Britain. He co-edited *Writing the Environment* (Zed Books, 1998), the first British collection of ecocritical essays. He was the first Chair of the UK branch of the Association for the Study of Literature and Environment (ASLE), and is now a member of the ASLE Executive Committee. In 1990 and 1991 he received the BBC Wildlife Award for Nature Writing. Currently, he is writing a monograph entitled *Beginning Ecocriticism* for Manchester University Press, and editing a new collection of British ecocritical essays for University Press of Virginia.

Jo Littler is Senior Lecturer at Middlesex University, London, UK. She is the author of *Radical Consumption? Shopping for Change in Contemporary Culture* (Open University Press, forthcoming, 2008); co-editor (with Roshi Naidoo) of *The Politics of Heritage: The Legacies of 'Race'* (Routledge, 2005) and editor of the 'Celebrity' issue of *Mediactive* (Lawrence and Wishart, 2004). She is also a regular book reviewer for *The Guardian* and reviews editor for *Soundings: A Journal of Politics and Culture.*

Martin Ryle is Reader in the Centre for Continuing Education and the English Department at the University of Sussex, where he teaches literature and cultural studies. His publications include *To Relish the Sublime?* (with Kate Soper, Verso, 2002) and *George Gissing: Voices of the Unclassed* (edited with Jenny Bourne Taylor, Ashgate, 2005). He is currently working on contemporary English and Irish fiction. He is a member of ASLE (Association for the Study of Literature and the Environment).

Roberta Sassatelli is Associate Professor of Cultural Sociology at the University of Milan. She has previously taught at the University of East Anglia (Norwich, UK) and the University of Bologna (Italy). She has been International Research Fellow within the ESRC/AHRC 'Cultures of Consumption' Programme. Her research focuses on the historical development of consumer societies and the theory of consumer action, with a particular interest in the contested development of the notion of the consumer as against notions such as citizen or person. She has done research on the commercialisation of sport and in particular on how fitness has been variously appropriated as a leisure pursuit. She has just completed research on how Italian consumers are responding to the food crisis and are articulating notions of typicality and quality, as well as on critical consumerism and alternative consumer practices. Her recent publications in English include *Consumer Culture: History, Theory and Politics* (Sage, London, 2007). Currently, she is completing a book on gym cultures for Palgrave.

Dr Gill Seyfang is Research Councils UK Academic Fellow in the School of Environmental Sciences at the University of East Anglia. Her research on 'Low Carbon Lifestyles' examines the social implications of moves towards more sustainable consumption patterns. She has an international reputation for groundbreaking research linking sustainability policy agendas with 'new economics' theories and cutting-edge community-based practice. Her work covers sustainable local development, the social economy, community currencies, local organic food systems and low-impact eco-housing.

Juliet Solomon (London Metropolitan University, Transport Research and Consultancy) is known on issues concerned with users, accessibility, young people, local transport policy, rural and community transport, as well as Green issues. She has made a number of contributions to the mainstream environmental debates, and in this context is probably best known as the author of *Green Parenting* (Macdonald Optima 1990). She coordinated the path-breaking study 'Social Exclusion and the Provision & Availability of Public Transport' for the DETR (2000). She is currently directing work on 'transport need' for the AUNT-SUE project (Accessibility and User Needs in Transport in Sustainable Urban Environments), which is part of a group of projects being funded and run by the EPSRC.

Kate Soper is Professor of Philosophy at the Institute for the Study of European Transformations at London Metropolitan University. Her most recent publications include *What is Nature? Culture, Politics and the Non-Human* (Blackwell, 1995, reprinted 1999, 2001, 2002) and *To Relish the Sublime?* (with Martin Ryle, Verso, 2002). She is a contributor to the *Women, Gender and Enlightenment* collection (edited by S. Knott and B. Taylor) recently published by Palgrave (2005). She was a lead researcher in the ESRC/AHRC 'Cultures of Consumption' Programme research project on 'Alternative Hedonism and the Theory and Politics of Consumption' and has recently edited a collection on *Citizenship and Consumption* with the Programme Director, Dr Frank Trentmann (Palgrave, forthcoming).

Dr Lyn Thomas is Deputy Director of the Institute for the Study of European Transformations at London Metropolitan University, and has taught French in the University since 1989. Her writings include *Annie Ernaux, an introduction to the writer and her audience* (Berg, 1999), *Fans, Feminisms and 'Quality' Media* (Routledge, 2002) and *Annie Ernaux, à la première personne* (Stock, 2005). She is a member of the Feminist Review Editorial Collective. She was responsible for the media study in the ESRC/AHRC 'Cultures of Consumption' Programme research project on 'Alternative Hedonism and the Theory and Politics of Consumption'. She is currently part of a team researching 'On-line listener engagement with BBC radio programming' in the BBC / AHRC Knowledge Exchange Scheme.

Introduction: The Mainstreaming of Counter-Consumerist Concern

Kate Soper

Until very recently, consumerism was seen as problematic only by some religious groups, small bands of 'greened' socialists and Green Party supporters, and the No Logo protesters within the anti-globalisation campaigns. Now it has moved to the centre of mainstream political concerns. Alarms over global warming, and more widespread acknowledgement of the human contribution to it, have played a major role in this. But there have been a number of other factors at work, not least the anxieties many now feel about the socially exploitative aspects as well as environmental consequences of the Euro-American lifestyle. It is, for some at least, disturbing to live in a world in which sweatshop conditions of labour satisfy the greed and ever more conspicuous consumption of the already very wealthy, and the gap between richest and poorest grows to inflammatory proportions. Ordinary moral revulsion against the suffering and injustice resulting from recent neo-liberal economic policies here combines with a growing sense of exposure to the potentially explosive political consequences of such yawning divisions of wealth.

To this we can add the evidence of a growing disquiet over the negative legacy of the consumerist lifestyle itself for consumers themselves. Mainstream media now register on a daily basis this new climate of disenchantment, with its concerns over the stress, pollution, ill-health, childhood obesity, car congestion, noise, excessive waste and loss of the 'arts of living' that are the unwanted by-products of consumerism. It is reflected, too, in the concerns of policy makers with the economic and social effects of the high-stress, fast-food lifestyle, and in the recent studies that have indicated that buying more does not bring greater happiness, and economic growth has no direct correlation with improved levels of well-being (index, 2006; Diener and Seligman, 2004; Frey and

Stutzer, 2002; Lane, 2001; Layard, 2005; The New Economics Foundation 'Happy Planet' and cf. Ekins, 2000; Jackson, 2004; Jackson and Marks, 1999; O'Neill, 2007; Purdy, 2005). As the UK government's independent watchdog on sustainable development sums it up in its recent report on 'Redefining Prosperity',

> ever since the ground-breaking work of Abraham Maslow and Manfred Max Neef, psychologists and alternative economists have set out to demonstrate that, far from there being any automatic increase in wellbeing for every increase in levels of consumption, much of our current consumption is turning out to be a very inadequate surrogate for meeting human needs in a more satisfying, durable way.
>
> (Sustainable Development Commission, 2003)

There is now, moreover, a body of research indicating that as people have woken up to the 'inadequate surrogacy' of their consumerist lifestyle, and opted for less materialistic values, so they have gained in happiness and well-being (Brown and Kasser, 2005; Kasser, 2002, 2007). And even those who are most committed to keeping us in the shopping malls, the corporate capitalist giants and their supportive governments, come close occasionally to acknowledging their vulnerability to such 'awakenings' and the vagaries of public spending that might ensue. One detected a sense of this, for example, in the calls following the Twin Towers attack, for us to commit to 'patriotic shopping' as a way of showing our support for the Western way of life: calls whose contradictory interference in our private market choices was at odds with the usual neo-liberal view of consumers as 'sovereign', and which said much about the dependency of corporate power on our continued loyalty to a consumerist way of life. There must surely also be an increasingly general (even if almost unspoken) understanding that the growth economy, with its reliance on continued material expansion and expenditure, is in contradiction with the restraints on consumption that are needed to meet agreed targets on carbon emissions. It is therefore not helpful, as the Sustainable Development Commission points out, for voters

> to be lulled into a sense of false security that the status quo ('growth at more or less all costs', on an exponential basis, despite the ever stronger scientific evidence that the ecological crisis is beginning to run away with us) is likely to remain the status quo for very much longer.

Yet here in Britain, we continue to live in a phase of extraordinary denial on this. The obsession with climate change and its apocalyptic scenarios runs in tandem with the constant encouragement to acquire ever more goods and services; government presses on with the expansion of motorways and airports even as it funds public information campaigns on the importance of reducing our personal carbon footprint.

In all these ways, then, and for all these reasons, the consumerist lifestyle is beginning to generate new tensions even within its own 'Western' heartlands, even as it continues to offer to less wealthy nations a virtually unchallenged model of progress and human prosperity.

'Alternative hedonism': A new approach to the critique of consumerism

It is these new forms of concern and conflictual development that have prompted the wide-ranging interdisciplinary perspectives on consumer culture that are represented in this volume. They have also provided the context for my own work on the concept of 'alternative hedonism' (that is, the pursuit and enjoyment of other pleasures) – a concept to which the title of this collection refers and to which its argument as a whole is ultimately related. The 'alternative hedonist' responds to the current situation not only as a crisis, and by no means only as presaging future gloom and doom, for she or he sees it also as an opportunity to advance beyond a mode of life that is not just unsustainable but also in many respects unpleasurable and self-denying. Alternative hedonists can speak more compellingly, and persuasively, than the prophets of environmental catastrophe. Whereas predictions of environmental disaster encourage a *carpe diem* fatalism, alternative hedonism is premised on the idea that even if consumerism were indefinitely sustainable it would not enhance human happiness and well-being (not, at any rate, beyond a point that we in the rich world have already passed). And it points to new forms of desire, rather than fears of ecological disaster, as the most likely motivating force in any shift towards a more sustainable economic order.

The general argument of alternative hedonism can be summarised in two main claims. The first, and more speculative, is that the affluent Euro-American mode of consumption, which has become the model of the 'good life' for so many other societies today, is unlikely to be checked in the absence of a seductive alternative conception of what it is to flourish and to enjoy a high standard of living. In this sense, the chances of developing or reverting to a more ecologically sustainable use of resources, as well as of doing away with some of the worst forms of social exploitation, are dependent on the emergence and embrace of new modes of thinking about human pleasure and self-realisation, especially, in the first instance, within the world's richer societies. This is not to suppose that ambivalence towards consumer culture will be experienced only by those who already have access to it; nor is it to presume that citizens of less affluent societies will necessarily be influenced by any alternative hedonist rethinking that might emerge within the more affluent. All that is claimed is that an important stimulus of any eventual change of direction will be the attractions for well-to-do consumers themselves of an alternative vision of the 'good life'. A counter-consumerist ethic and politics should therefore appeal not only to altruistic compassion and environmental concern, but also to the more self-regarding gratifications of consuming differently. It should develop and communicate a new erotics of consumption or hedonist 'imaginary'.

The second and more substantive claim is that we are now, as already noted, seeing the beginnings of such a trend, both in the sense that new conceptions of the good life appear to be gaining a hold among some affluent consumers and in the sense that there is a more questioning attitude towards the supposed blessings of consumerism (Bunting, 2004; Hodgkinson, 2004; Levett, 2003; Purdy, 2005; Schor, 1991; Shah, 2005; Thomas, 2008). People are beginning to see the pleasures of affluence both as compromised by their negative effects and as pre-empting other enjoyments. The enjoyment of previously unquestioned activities – such as driving, or flying, or eating out-of-season strawberries that have been transported halfway round the world, or buying a new refrigerator – is now tainted by a sense of their side-effects. The pleasures of the consumerist lifestyle as a whole are troubled by an intuition of the other pleasures that it constrains or destroys, especially those that would follow from a slower, less work-dominated pace of life.

Alternative hedonism represents a critical approach to contemporary consumer culture that is distinctive in its concern with self-interested rather than altruistic motives for shifting to greener lifestyles. The citizen and subject invoked here is rather more complex, and lifelike, than the narrowly appetitive individual imagined by neo-classical economics and rational choice theory. The individual motivated by alternative hedonism acts with an eye to the potentially negative impact of aggregated personal acts of affluent consumption for consumers themselves, and takes measures to avoid contributing to it. For example, he or she decides to cycle or walk whenever possible, in order not to add to the pollution, noise and congestion of car use. However, the hedonist aspect of this shift in consumption practice does not lie only in the desire to avoid or limit the unpleasurable by-products of collective affluence, but also in the sensual pleasures of consuming differently (Levett, 2003, 60f). There are intrinsic pleasures in walking or cycling which the car driver will not be experiencing. But cycling or walking themselves are much pleasanter, and may only be possible, where car use is limited – that is, where others too are making alternative hedonist commitments to self-policing in car use and are supporting policies that restrain it. In this sense, the more selfishly motivated preference for cycling will be inseparable from a more collectively oriented concern to avoid contributing to noise, danger, pollution and congestion. Similarly, those who avoid fast food are likely to do so for a complex of more or less self-interested motives, since to be bothered about its impact on one's own health is usually also to be bothered about the processes of manufacture. But the pleasure of eating healthy food may well be enhanced by the altruistic or moral pleasure of knowing that one has also avoided contributing to certain forms of environmental destruction and social exploitation. There is clearly, then, a considerable overlap between alternative hedonist types of motivation and the altruism of the green or ethical consumer. Both might be said to engage in a distinctively moral form of self-pleasuring and in a self-interested form of altruism which takes pleasure in committing to a more socially accountable mode of consuming.

The concept of alternative hedonism highlights this complexity of desire and motivation in newly emergent consumer responses. In dwelling on the pleasures of escaping the consumerist lifestyle rather than on the need for frugal consumption, its emphasis differs from that of much of the literature on ethical consumption and sustainable development. Yet it clearly chimes with those calling for a redefinition of prosperity (Evans and Jackson, 2007; Jackson, 2004, 2006; Kasser, 2007; Sustainable Development Commission Report, 2003) and with the growing demands for the GNP measure of productivity to yield to others – such as the UNDP Human Development Index (HDI) and the Index of Sustainable Economic Welfare (ISEW) – that are more reflective of real levels of well-being rather than purely quantitative economic growth. (The UK government's 'Quality of Life' index is a welcome response to such demands and a step in the right direction although its profile to date remains very low.)

Alternative hedonism also offers a theoretical framework for reflecting on the reasons behind the formation of the growing number of campaigning networks linking those who have opted for 'down-shifting', reduced working hours and more sustainable lifestyles. The 'Voluntary Simplicity' movement in the United States aims, for example, to promote a way of living that is 'outwardly simple, inwardly rich' (www.simpleliving.net/; cf. Elgin, 1993; Holst, 2007; Pierce, 2000), while the mission of the more recently formed Center for the 'New American Dream' is 'to help people live the dream, but in a way that ensures a livable planet for current and future generations'. Insisting that its message is not about deprivation but about getting more of what really matters – 'more time, more nature, more fairness, and more fun' – the Center can fairly claim to be attempting 'nothing less than a shift of American culture away from an emphasis on unconscious consumption towards a more fulfilling, just, and sustainable way of life' (www.newdream.org). To this we may add the growing expressions of dissent from the work-driven society, and the new interest in 'time affluence', that are being registered in the United States and across Europe (Bunting, 2004; de Graaf, 2003; Kasser, 2007; Schor, 1991), and the continued expansion of the 'Slow Food' (www. slowfood.com; www.slowfood.org.uk) and 'Slow City' (www.cittaslow.net; www.cittaslow.org.uk) networks (cf. Honore, 2005).

By focussing on these new developments and shifts of feeling in constituting an immanent critique of consumer culture, the 'alternative hedonism' perspective aims to avoid the moralising about 'real' needs that has often characterised earlier critiques of consumer culture (Miller, 2001). It engages with ambivalence or disaffection towards consumerism as this comes to the surface and finds expression in the sensibility or behaviour of consumers themselves. Although these shifts in response, and the new representations of pleasure that go with them, are presented in a positive light, the primary aim is not to defend or justify certain forms of consumption as objectively more needed or authentic. The concern is not to prove that consumers 'really' need something quite other than what they profess to need (or want) – a procedure which is paternalistic and undemocratic – but to reflect on the hedonist aspirations prompting changes in experienced or imagined need and their implications for the development of more sustainable modes of consumption. This approach marks a break in the politics of 'counter-consumerism' from more orthodox left responses to the consumer society.

Counter-consumerist critique: Historical precursors and contemporary perspectives

The concept of alternative hedonism elaborated above has been developed in a recent research project (2004–2006) undertaken by myself and Lyn Thomas in the ESRC/AHRC funded 'Cultures of Consumption' Programme. The theoretical refinement of the arguments was here anchored to a media study carried out by Thomas (whose findings she draws upon in her chapter in this volume). This was explicitly designed to highlight the ways in which the ideas and practices addressed in the theory are now moving out of the counter-cultural margins and finding a register in mainstream news media, lifestyle magazines and popular TV programmes. It allowed the argument to draw on indices of interest in alternative hedonism now manifest in the culture at large and entering into everyday experience. Of course, this interest could well be deflected into new forms of niche marketing, rather than inaugurating any serious challenge to the current order. It remains to be seen where it will lead, and how, if at all, it will connect with any more radical counter-consumerist politics. But those in the academy, and in public life generally, who would wish to see it develop into a more explicit and challenging form of opposition have a responsibility to engage with its tensions, reflect upon its ethical and aesthetic implications, and provide more explicit cultural representation of the non-puritanical but at the same anti-consumerist 'political imaginary' to which it is gesturing.

The contributors to this volume have all in differing ways taken up that responsibility. They engage with the counter-consumerist cultural and political turn by reconsidering some established approaches to consumer society, by discussing forms of pleasure and happiness that challenge consumerism's priorities, by criticising and deconstructing its imaginary representations, and through philosophical reflection on underlying values. Some contributors engage directly with the alternative hedonist framework of thinking: several chapters began life as papers presented to a conference in April 2006 at London Metropolitan University, organised by Thomas and myself as part of our research project. In the book as a whole, the reader will find a range of disciplinary approaches, topics of engagement and points of cultural reference, reflecting the authors' diverse backgrounds and geographical locations. Although Britain is the main reference for most of the chapters, the economic and cultural questions addressed have a much wider temporal and spatial relevance, bearing as they do on the global impacts, human and ecological, of the pursuit of the consumerist way of life. The collection includes contributions from writers in the United States and countries in continental Europe, and relates, as indicated above, to a range of international counter-consumerist critiques and campaigns.

Despite their differences of approach and origin, the contributors are united by a shared sense of the importance of addressing questions about the good life, pleasure and self-fulfilment in relation to consumption. They represent a growing interest in developing a new critical understanding of consumerism within the humanities. As the range of perspectives brought to bear suggests – and demonstrates – there is a clear need for a broad inter-disciplinary understanding of how the consumer society and its possible future is theorised, researched and taught in higher education. In providing a normatively accented and multi-perspectival engagement of that kind, this collection differs from those in cultural studies and sociology that deal with contemporary consumer culture in a more purely descriptive or value-neutral manner. In virtue of their commitments, several of the contributors address, for example, the conditions and forms of agency that might help over time to bring about a fairer allocation and more responsible and life-enhancing use of global resources. Their engagement as academics with the fracturing of the consensus about the good life and the practices associated with it is in this sense not purely 'academic', but follows from their interest in whether and how new forms of thinking and representation are registering discontent or beginning to influence a shift to a more enjoyable, socially just and environmentally sustainable consumption. In these respects, their studies share interests and values in common with those of the environmental and anti-globalisation movements, green politics and the campaigns for Fair Trade and ethical consumption. Such alignments, the editors would argue, are those that an imaginative and socially accountable academic engagement with consumption in the era of accelerating climate change should actively be seeking to cement. Moreover, to ignore the emergent new structures of feeling within consumer culture is to abstract from some of its most dynamic, interesting and politically significant dimensions. As Ulrich Beck has argued in his criticism of positivism in sociology, the discipline

> loses credibility if it walls itself off against these social movements and perceptions that urge reform, and therefore criticism, of industrial society. Every skilled industrial worker, every high school graduate, every conservative with what has now become a quite normal ecological awareness has concrete possibilities of social change which a sociologist as analyst of this society is forbidden by the established

> super-ego of his profession from even registering... 'Reflexive modernization' means that society itself produces multiple and concealed forms of self-criticism which cannot be perceived and decoded by a sociology that abstains from social criticism.
>
> (Beck, 1996, p. 175)

Critical engagement today with the consumer society is not, of course, without precedents, and clearly connects with an earlier left-wing tradition of Marxist and Frankfurt School critique of commodification and 'commodity aesthetics'. Yet there are also definite differences of emphasis and outlook, beginning with the attention paid in the arguments represented here to the domain of consumption as a potential source of ethical pressure and political agency. For although Marxists and those influenced by Critical Theory drew attention to the negative impact of the market on human fulfilment, they presented this primarily in terms of its 'false' forms of provision and theft of consumer autonomy. The emphasis was on the construction or manipulation of consumer 'needs' and wants (Adorno, 1974, 1991; Bauman 1988, 1991, 1998; Gorz, 1989; Haug, 1986 and 2006; Jameson, 1990; Leiss, 1978; Lodziak, 1995; Marcuse, 1964, 1972), rather than on the critical reflexivity of consumers; and production alone was seen as the site of potential mobilisation against the capitalist order, through the agency of worker militancy. The market society, on this view, had protracted its domination by subverting the will to resist it or to enjoy any system of pleasures other than the one it provided. Resistance was thus theorised as prompted solely by the exploitations of the workplace, while consumption, viewed as essentially determined and controlled by production, was regarded as exercising a placating influence and as tending to reconcile consumers to the existing order rather than firing them to oppose it. This politics, moreover, in its most orthodox form, was directed primarily at transforming the relationships of ownership and control over industrial production rather than at the quality and methods of production as such; it was about equalising access to consumption, rather than revolutionising its culture.

This abstraction from the politics of consumption first began to change with the emergence of 'Flower Power' and the new social movement campaigns of the 1960s. But it was really only in the arguments of the 'Red–Green' formation, as theorised initially and most influentially by Rudolf Bahro, that some parts of the left associated themselves more closely with a distinctive 'counter-consumerist' critique (see Bahro, 1982, 1984; Benton, 1991; O'Connor, 1998; Ryle, 1988; Redclift and

Woodgate, 1995; Soper, 1991; Williams, 1983, 1993). As one of the editors of this book put it at the time, 'a green "politics of consumption" or' life-style politics is developing, which calls on us to accept 'that "the consumer" (each' one of us) is something more than a passive victim of the capitalist expansion of needs' (Ryle, 1988, pp. 90f.). Much of this red–green literature also emphasised the success of capitalism in enhancing the material prosperity of the working class to a point where it made little sense to continue to invoke it in theory as the sole possible agent of political opposition to the existing economic order.

Ulrich Beck and others have since then theorised the shift that has taken place from a working-class politics organised around provision for basic material needs to a trans-class politics of 'risk' organised round the fears of contemporary consumers (Beck's tendency has been to emphasise our collective victimisation by industrial pollution rather than our collective implication in its creation) (Beck, 1992; cf. Giddens, 1991, pp. 109ff.). In this situation, labour militancy and trade union activity have largely become confined to protection of income and employees' rights within the existing structures of globalised capital, and have not sought to challenge, let alone transform, the 'work and spend' dynamic of affluent cultures. It is not, one may therefore argue, at the level of production or through the agency of the working class that any transformative pressure of that kind is presently likely to arise. Should it do so, it is more likely to be in the realm of consumption – in decisions to simplify the way we live and settle for a less materially encumbered and work-driven existence. This would involve deciding *not* to buy, but to do without; deciding to boycott the commodities of the global brand names, or just leave them on the shelf; deciding to avoid supermarkets and shopping malls and to purchase and invest only in goods and services of proven ethical and green credentials. Such 'agency' would no longer be class specific, but more diffusely exercised – even if in the first instance many of the more rebellious consumers would probably be relatively well off. This kind of counter-consumerist political perspective thus marks a significant departure from earlier critiques of capitalism, although it certainly recognises the inter-dependence of production and consumption.

Such a perspective also differs from neo-liberal, orthodox left-wing and postmodernist understandings of consumer formation and agency, in that it regards choice neither as wholly autonomous and freely adopted, on the one hand, nor as wholly formed (manipulated or constructed) by trans-individual systemic pressures, on the other. Its focus, rather, is on the relative autonomy of the ways in which people are now beginning to reflect upon their own consumerist formation in the light of its negative consequences both personally and collectively. It thus seeks to develop the complex and nuanced theoretical understanding required to accommodate the troubled and equivocal forms of contemporary consumer reaction to consumerism. As some of the chapters in this volume indicate, these reactions only come into being as a consequence of immersion in consumer culture. Along with their related ethical, aesthetic and environmental concerns, they are thoroughly conditioned by this context, and shaped by the media and other sources of information. Yet the decision to act on this dissatisfaction, the supra-personal level of concern for the collective impact of aggregated consumer choices and the high degree of reflexivity involved in counter-consumerist reconceptions of the good life all indicate a degree of autonomy and accountability of which we also need to take account.

Another distinctive aspect of several of the chapters lies in their emphasis on the hedonist rewards, rather than the duties, of living and consuming differently. While critical of the dystopian aspects of modern affluence, and sceptical about many of its supposed pleasures, they do not echo the puritanical tone of some earlier left-wing critiques on consumerism. They neither endorse consumerist forms of pleasure-seeking and fulfilment, nor pronounce jeremiads against their excesses and self-indulgence. Contributors here write from a position that acknowledges the extent to which we are all currently party to consumerism, but goes on to emphasise the displeasures of the consumerist lifestyle and the fulfilments, both sensual and spiritual, that we might otherwise be able to enjoy.

The theorists of ethical consumption and 'virtuous shopping' have made comparable points about the reflexivity and accountability of many contemporary consumers (Barnett and others, 2005; Harrison, Newholm and Shaw, 2005; Micheletti, 2003; Micheletti and Peretti, 2003; Miller, 1995). This approach, with its implicit critique of the model of the supposedly wholly self-seeking consumer and its attention to more 'republican' aspects of consuming, also has affinities with recent writings on the 'citizen-consumer' (Daunton and Hilton, 2001; Ginsborg, 2005; *Journal of Consumer Culture,* 2007; Sassatelli, 2007; Soper and Trentmann, 2007; Trentmann, 2006). The emphasis on the complex and contradictory gratifications and costs of consumption differentiates such perspectives from those of much recent consumption theory. A too exclusively semiotic – and often rather celebratory – preoccupation with fashion, self-styling and identity-affirming forms of consumption has characterised much postmodernist engagement with consumer culture. There are, of course, narcissistic satisfactions afforded by buying things, and many people clearly do just love shopping and the sense of belonging and forms of self-assurance provided by fashion articles and brand logos. But these postmodernist moves away from an earlier account of tastes and choices in consumption as a relatively predetermined expression of class position (along the lines classically formulated by Bourdieu) have surely made too much of the idea that one can individualise oneself by shopping. As Alan Warde has argued, the liberating aspects of choice have been exaggerated: although product differentiation may be imperative for profit, the effect 'is not highly distinctive. Extensive variety encourages undistinguished difference. The world of consumption is led less by great personal aesthetic imagination, more by the logic of the retailing of commodities' (Warde, 1997, pp. 194, 201ff). On the other hand, from a counter-consumerist point of view there are problems with the relatively uncritical account that Warde and others give of the less self-centred routine practices of everyday consumption (Schatzki, 1996; Schatzki *et al*, 2001; Warde, 1997; 2004). For it is precisely certain mundane forms of everyday consumption, such as driving and flying, which are responsible for some of the worst environmental damage – and which are now generating negative by-products for consumers themselves.

The focus associated with 'alternative hedonism' is neither on consumption as a bid for personal distinction or individualisation nor on consumption as a relatively unconscious 'form of life', but on the ways in which a whole range of contemporary consumer practices, both more or less 'everyday' and more or less identity-oriented, are being brought into question by reason of their environmental consequences, their impact on health and their distraints on both sensual enjoyment and more spiritual forms of well-being.

This collection

In the foregoing, I have introduced and contextualised the arguments associated with the research on 'alternative hedonism' that provided the initial prompt for this volume. In doing so I have noted some common points of reference and outlook. I have also indicated the wide range of perspectives and commitments represented here, and the chapter notes that follow are designed to provide an amplified sense of this.

The first of the three parts into which the collection is divided, on 'Texts and Representations', brings together writings that represent a relatively new kind of engagement in media and cultural studies. This approach reflects critically on the examples of contemporary media and culture in their various and contradictory roles as market adjunct and purveyor of consumerist fantasies, as register of new forms of equivocation about consumption, and as resources for an alternative hedonist and counter-consumerist 'imaginary'. Read together, the chapters provide a complex of dialectical insights into the ways in which current representations can collude in the reproduction of consumer culture; can reflect, even if only implicitly, emerging forms of disenchantment with it; or can offer utopian pointers to other futures.

In the opening chapter, Roberta Sassatelli draws attention to the shifts in the role of differing agencies (mainstream advertising, consumer protection organisations, critical consumption movements, counter-advertising) in representing consumers themselves, thereby providing them with their diverse narrative identities. In a context in which capitalism has itself become increasingly attuned to the cultural critique of consumerism, the symbolic boundaries defining 'the consumer' are destabilised, and the latter emerges as a site of contestation between, on the one hand, the systemic pressures of commercialism, and, on the other, the agencies seeking to promote, and to encourage consumers themselves to adopt, a more ethical and socially accountable conception of the role of the contemporary shopper. We need, therefore, to acknowledge the extent to which countering discourses disturb the usual construction of the consumer as an uncritical participant in contemporary society. Though the consumer remains an equivocal subject category, the antinomy between commercial and ethical aims may yet figure, Sassatelli suggests, as a dialectical resource to modify views on the market economy.

Martin Ryle's chapter is concerned primarily with representations in contemporary fiction that invite critical scrutiny of consumer culture. Linking his argument directly to the 'alternative hedonist' critique, Ryle draws attention to the pleasures lost or pre-empted in the advance of consumerism and defends the contemporary political relevance of a qualified nostalgia for these vanished sources of delight and satisfaction. Although the antinomies of destructive production may have been repressed, they are now returning, he suggests, to haunt the ways in which we understand both progress and retrospect. Referring us to Raymond Williams' seminal discussion of 'retrospective radicalism' in the *Country and the City* and his prescient warning against glib dismissals of this (Williams, 1993), Ryle reveals the ways in which a 'retrospective' sensibility, critical of the dystopian 'progressive thrust' of consumer culture, is at work in recent novels by John McGahern, Michel Houellebecq, Kazuo Ishiguro and Ali Smith.

Lyn Thomas, in her chapter, builds on earlier research on expressions of ambivalence about the work-dominated, consumerist mode of living to be found in recent popular television. She considers how far and in what ways this ambivalence and its associated yearnings are now registered in the even more commercially driven, and hence presumptively less hospitable, medium of lifestyle magazines. She points to the ways in which this type of coverage is serving the ends of new forms of niche marketing or corporate green-washing (often enough of the palest hue). But her engagement also sheds light on the real and rather sad forms of subservience to the consumerist lifestyle, and the desires to escape its confines, that are presumed in the readers of these journals. What is of interest here, as she notes, is the expression, in the midst of images of plenty, of a sense of dissatisfaction with the stresses and strains of lives devoted to earning and consuming and of the need to escape those pressures.

Though seriously doubting the radical import of the coverage she charts, Thomas draws our attention to the significance of the socio-cultural shift that has taken place when a Tesco promotional magazine must seek to reposition the company as a saviour of the planet. Simon Blanchard, by contrast, is uncompromisingly dismissive of the counter-consumerist potentials of the mainstream television engagements with environmental issues that he discusses in his chapter. Acknowledging the seeming tension between the consumerist offering of most TV schedules and occasional media outbreaks of 'caring and sharing', such as the G8/'Live 8' saga, he argues that the latter, by keeping within the framework of neo-liberal critiques and 'solutions', in fact serve only as a means to reinforce a sense of human 'exemption' from natural limits and constraints. We should therefore be under no illusions about the reluctance of mainstream TV to contest consumerism. With the ascendancy of the post-1970s neo-liberal 'telegentsia' that has overseen the marketisation of TV's institutional structure, television, he claims, has become ever more collusive in the growth of consumer society and the populist-authoritarian and corporate regime on which it now thrives. For a register of any more serious challenge to consumerism, we must for the moment look to more marginal media: websites, festivals, the documentary renaissance outside TV.

The second part, 'Value, hedonism, critique', brings together chapters that address the more philosophical issues raised by a counter-consumerist hedonism, engaging with its historical precursors and influences, with the dialectics of the 'alternative hedonist' form of response to instrumental rationality and with the understanding and analysis of the forms of life-politics registered in current disquiet and disaffection with consumer culture. In a chapter oriented towards American responses to consumerism and the mobilisation of an anti-consumerist social movement, Sam Binkley points to some limitations in current interpretations based on theories of growing consumer self-reflexivity and concern for a life-politics and turns to Bourdieu's concept of 'habitus' as a more adequate theoretical tool for grasping the character of the pre-reflexive and unconscious dispositions at work in recent responses to consumerism. What this reveals, he suggests, is a 'new bohemianism' that, although itself an embedded dynamic of a post-Fordist economic order and in many ways complicit in it, is also driven by an underlying logic at odds with the instrumental rationality and inauthentic modes of mass market provision. While this falls well short of a sustained political strategy, Binkley notes its potential importance as a basis for generating a more reflexive awareness of the daily practices of shopping and spending. He suggests that as this avant-garde 'habitus' extends its influence, more people are likely to be inflected with a specifically bohemian sense of aesthetics in daily life and to develop an implicit opposition to a market system short on corporeal pleasures and expressive embodiment. A reflexive self-awareness that is part and parcel of sub-politics and life politics might thus move from being an explicit reflexive discourse to being an internalised mode of daily practice more common to all.

In his chapter, Marius de Geus discusses the ways in which contemporary concerns over sustainable consumption have revived interest in classic philosophical questions about the quality of the 'good life', how it can be responsibly pursued and what is meant by the 'art of living'. His chapter engages with this new interest, exploring the relevance to contemporary concerns and controversies of the two main traditions of thinking on the 'art of life', that of the Aristoteleian 'moralists' who advocate the rewards of frugality and moderation and that of the 'hedonists' who emphasise the importance of pleasure and the role of luxury and expanding consumption in securing it. Recognising the inadequacies of both traditions to meet the demands of our times for lifestyles that are at once self-chosen, sustainable and enjoyable, he argues for an alternative hedonism that integrates respect for individual freedom with social-ecological and moral accountability. Such an approach, he suggests, should be based on pragmatism, respect for the diversity of lifestyles compatible with living within ecological limits, feelings of responsibility towards the present and future generations, and the basic empirical insight that abundance does not make people happier.

The importance of reconciling tensions between human and environmental demands is also a main concern of Richard Kerridge's chapter. In line with the 'alternative hedonist' emphasis on the role of desire in promoting sustainable consumption, Kerridge here argues that environmentalism should hold out a vision of pleasure, rather than exhorting people to self-deprivation. Could we, he asks, act to avert environmental crisis not only because we have been persuaded of the need to do so, but because our impulses and appetites come to lead us in that direction? Kerridge looks in particular to recent nature writing and eco-phenomenology as potential resources for such a project, but does so in ways that are mindful of the paradoxical clashes between the two aspects of Romantic yearning often registered in this writing: on the one hand, that of immanence and self-forgetting, of the return to nature and absorption in the moment; on the other, that of gazing from outside, of the anticipation and memory associated with fantasies of who we are or might become. He also acknowledges that any ecocritical recourse to Romanticism might be thought problematic in terms of Colin Campbell's influential view that Romantic fantasy and desire themselves underpin the development of consumerism, rather than acting as a restraint upon it. Kerridge's overall position is that a hedonism that can both respond to the environmental needs of our times and prove adequate to the complex aspirations of those formed within contemporary industrialised culture would need to accommodate the tensions between ironic distance from nature and the yearning for reunion with it. Whether this can come about in time is doubtful: Kerridge insists on the imminent dangers posed by climate change. But if Romanticism is still latent within consumerism, as even Campbell's argument implies, might this not now be renewed in non-consumerist forms and promote some bridging or synthesis of pre- and post- industrial sensibilities?

Contributions to the final part, on 'Everyday consumption', address issues relating to the policy and practices of everyday life and consider how academic engagements with consumption should be updated and rethought in the light of current political developments. Contributors discuss the links between counter-consumerist campaigning and other social movement agendas, and reflect on strategies for promoting sustainable forms of consumption.

Theorists and critics of consumer culture and its seductions often focus on material consumption without giving due attention to the complex of pleasures and motivations at play in the satisfactions it offers. Juliet Solomon offers a corrective to this in her chapter on the human need for mobility. She discusses the evidence which indicates that travel – in all its various modes, whether sustainable or otherwise – provides intrinsic pleasures, both sensuous and psychological, in addition to the instrumental benefits of travelling from one point to another. Solomon considers how movement, even commuting, has become a significant part of the spectrum of human pleasures, and reflects on the implications of a reduced consumption of mobility for our quality of life and the need for meaningful substitutes. She suggests that the arguments for sustainable transport will be all the more compelling if they acknowledge the fundamental need of movement for movement's sake and the differing ways in which this can be enjoyed.

Jo Littler's chapter moves from reflection on the personal resonance of a spur-of-the-moment purchase (of an anti-Starbucks badge) into a richly nuanced engagement with the connections between femininity, gender and the critique of consumerism. Against the background of a too exclusive engagement in cultural studies with ways that mass consumer culture has been gendered as 'feminine', Littler argues that academics need to 'catch up' with new strands of popular politics by revisiting the relations between consumption and feminism, and herself offers new genealogies for studies that are at once both *pro*-feminist and *anti*-consumerist. One route into this, she suggests, is via the history of female involvement in 'political consumerist' campaigning; but given the reliance of this on the traditional association of women with the domestic sphere, we need also to open up a framework of thinking in which essentialist presumptions of this kind can be challenged, and to provide an altogether wider repertoire of gendered possibilities of participation in, and critique of, consumption. This should now be robust enough to encompass criticism of women's co-option in consumer inequality and over-consumption.

Aligning herself with those who are critical of mainstream policy approaches to sustainable consumption for decoupling economic growth from environmental degradation, Gill Seyfang in her chapter reports on research designed to test the practical social implications of the alternative favoured by theorists of the 'New Economics'. (The research was carried out as part of the Programme on Environmental Decision-Making funded by the Economic and Social Research Council, and the chapter is a revised version of a journal article: details are in Seyfang's Acknowledgements.) Empirical evaluation of a local organic food network in Norfolk shows how successfully this measures up against the five key criteria of the 'New Economics' approach (localisation, ecological footprint reduction on the part of developed countries, community-building, collective action and the creation of new socio-economic institutions). The reported interviews with consumers are revealing of the 'alternative hedonist' complex of motivations and responses to local and organic food provision, and the study also interestingly suggests that organic food is less a preserve of a rich élite than is often supposed.

As the range and diversity of its discussions indicate, this collection engages directly with the tensions resulting from the dependency of the globalised economy on the promotion of a consumerist way of life which is at once closely associated with 'freedom and democracy' and at the same time socially and ecologically damaging. It thus challenges the limitations of orthodox approaches to consumer society and brings a number of unusual optics to bear on its critique, especially from within the domain of cultural politics. It is also distinguished by its multi-disciplinary perspective on the consumer, its focus on the more pleasurable dimensions of sustainable consumption and the attention it pays to questions of ethics, aesthetics, citizenship and environment. For all these reasons, it is hoped that the volume will provide and provoke informed reflection on public discourses, representations, policies and debates. It aims to intervene in those debates, to offer models (to students and general readers) of accessible and stimulating critical-cultural enquiry based on reflection and research and to stimulate and encourage further interdisciplinary and collaborative work on consumption and alternative hedonism

Acknowledgements

I thank my co-editors, the participants in the 'Counter-consumerism: religious and secular perspectives' conference at London Metropolitan University, April 2006, and colleagues in ISET, at London Metropolitan University, for their support.

I acknowledge the support of Award no. RES-154-25-005 in the ESRC/AHRC Programme on the 'Cultures of Consumption' for support on research (with Lyn Thomas) on 'Alternative hedonism and the theory and politics of consumption'.

References

Adorno, Theodor (trans. E.F.N. Jephcott) (1974; first published 1951) *Minima Moralia: Reflections from Damaged Life* London: New Left Books.

Adorno, Theodor (ed. Jay Bernstein) (1991) *The Culture Industry: Selected Essays on Mass Culture* London: Routledge.

Bahro, Rudolf (trans. David Fernbach) (1982) *Socialism and Survival* London: Heretic Books.
Bahro, Rudolf (trans. Gus Fagan and Richard Hurst) (1984) *From Red to Green: Interviews with New Left Review* London: Verso.
Barnett, Clive, Cloke, Paul, Clarke, Nick and Malpass, Alice (2005) 'Consuming Ethics: Articulating the Subjects and Spaces of Ethical Consumption' *Antipode,* 37(1), pp. 23–45.
Bauman, Zygmunt (1988) *Freedom* Milton Keynes: Open University Press.
Bauman, Zygmunt (1991) *Modernity and Ambivalence* Cambridge: Polity.
Bauman, Zygmunt (1998) *Work, Consumerism and the New Poor* Milton Keynes: Open University Press.
Beck, Ulrich (trans. Mark Ritter) (1992) *Risk Society: Towards a New Modernity* London: Sage.
Beck, Ulrich (1996) *The Reinvention of Politics: Rethinking Modernity in the Global Social Order* Cambridge: Polity Press.
Benton, Ted (1991) 'The malthusian challenge: Ecology, natural limits and human emancipation', in Peter Osborne (ed.) *Socialism and the Limits of Liberalism* London: Verso.
Brown, K. W. and Kasser, Tim (2005) 'Are psychological and ecological well-being compatible? The role of values, mindfulness, and lifestyle' *Social Indicators Research, 74*, pp. 349–368.
Bunting, Madeleine (2004) *Willing Slaves: How the Overwork Culture is Ruling Our Lives* London: Harper Collins.
Daunton, Martin and Hilton, Mathew (2001) (eds) *The Politics of Consumption: Material Culture and Citizenship in Europe and America* Berg: Oxford.
Diener, Ed and Seligman, Martin (2004) 'Beyond money: Toward an economy of well-being' *Psychological Science in the Public Interest, 5*, pp. 1–31.
Ekins, Paul (2000) *Economic Growth and Environmental Sustainability* London: Routledge.
Elgin, Duane (1993) *Voluntary Simplicity* (revised edition) NY: William Morrow.
Evans, David and Jackson, Tim (2007) 'Towards a Sociology of Sustainable Lifestyles' RESOLVE Working Papers 03–07.
Frey, Bruno and Stutzer, Alois (2002) *Happiness and Economics* Princeton: Princeton University Press.
Giddens, Anthony (1991) *Modernity and Self-Identity* Cambridge: Polity.
Ginsborg, Paul (2005) *The Politics of Everyday Life* New Delhi: Penguin Books.
Gorz, André (1989) *Critique of Economic Reason* London: Verso.
de Graaf, John (ed.) (2003) *Take Back your Time: Fighting Overwork and Time Poverty in America* San Francisco: Berret-Koehler.
Harrison, Rob, Newholm, Terry and Shaw, Deirdre (eds) (2005) *The Ethical Consumer* London: Sage.
Haug, Wolfgang (trans. Robert Bock) (1986) *Critique of Commodity Aesthetics: Appearance, Sexuality and Advertising in Capitalist Society* Cambridge: Polity.
Haug, Wolfgang (2006) 'Commodity aesthetics revisited' *Radical Philosophy*, 135, pp. 18–24.
Hodgkinson, Tom (2004) *How to be Idle* London: Hamish Hamilton.
Holst, Carol (ed.) (2007) *Get Satisfied: How Twenty People Like You found the Satisfaction of Enough* Wesport Connecticut: Easton Studio Press.
Honore, Carl (2005) *In Praise of Slowness: Challenging the Cult of Speed* New York: Harper One.

Jackson, Tim (2004) *Chasing Progress: Beyond Measuring Economic Growth* London: New Economics Foundation.
Jackson, Tim (2006) *Earthscan Reader in Sustainable Consumption* London: Earthscan.
Jackson, Tim and Marks, Nic (1999) 'Consumption, sustainable welfare and human needs' *Ecological Economics*, 28, pp. 421–442.
Jameson, Frederick (1990) *Late Marxism: Adorno or the Persistence of the Dialectic* London: Verso.
Journal of Consumer Culture (2007) Special issue on 'citizenship and consumption', 7 (2).
Kasser, Tim (2002) *The High Price of Materialism* Cambridge, MA: MIT Press.
Kasser, Tim (2007) 'Values and Prosperity', Paper to Sustainable Development Commission seminar on 'Visions of Prosperity', 26th November.
Lane, Robert (2001) *The Loss of Happiness in Market Democracies* Yale: Yale University Press.
Layard, Richard (2005) *Happiness: Lessons from a New Science* London: Allen Lane.
O'Connor, James (1998) *Natural Causes: Essays in Ecological Marxism* London: Guilford Press.
Leiss, William (1978) *The Limits to Satisfaction: On Needs and Commodities* London: Marion Boyars.
Levett, Roger (2003) *A Better Choice of Choice: Quality of Life, Consumption and Economic Growth* London: Fabian Society.
Lodziak, Conrad (1995) *Manipulating Needs, Capitalism and Culture* London: Pluto.
Marcuse, Herbert (1964) *One-Dimensional Man* London: Beacon Press.
Marcuse, Herbert (1972; first published 1955) *Eros and Civilisation* London: Abacus.
Micheletti, Michelle (2003) *Political Virtue and Shopping* New York & London: Palgrave, Macmillan.
Micheletti, M. and Jonha Peretti (2003) 'The nike sweatshop Email: Political consumerism, internet and cultural jamming', in M. Micheletti and others (eds) *Politics, Products and Markets: Exploring Political Consumerism Past and Present* New Brunswick: New Jersey.
Miller, Daniel (1995) 'Consumption as the vanguard of history', in Miller (ed.) *Acknowledging Consumption* London & New York: Routledge.
Miller, Daniel (2001) 'The poverty of morality' *Journal of Consumer Culture*, 1(2), pp. 235–243.
New Economic Foundation (2006) 'Happy Planet Index' www.happyplanetindex.org/reveals.htm (consulted 7 August 2007).
O'Neill, John (2007) 'Sustainability, well-being and consumption: The limits to hedonic approaches', in Kate Soper and Frank Trentmann (eds) *Citizenship and Consumption* London: Palgrave .
Pierce, L. B. (2000) *Choosing Simplicity: Real People Finding Peace and Fulfillment in a Complex World* Carmel, CA: Gallagher Press.
Purdy, David (2005) 'Human happiness and the stationary state' *Soundings*, 31, pp. 133–146.
Redclift, Michael and Woodgate, Graham (1995) *The Sociology of the Environment* Vol. 1, Part II 'Marxism and the Environment', Aldershot: Edward Elgar, pp. 253–606.
Ryle, Martin (1988) *Ecology and Socialism* London: Random House.

Sassatelli, Roberta (2007) *Consumer Culture. History, Theory and Politics* Sage: London.
Schatzki, Theodore (1996) *Social Practices: A Wittgensteinian Approach to Human Activity and the Social* Oxford: Oxford University Press.
Schatzki, Theodore, Cetina, Karin Knorr and von Savigny, Eike (eds) (2001) *The Practice Turn in Contemporary Theory* London: Routledge.
Schor, Juliet (1991) *The Overworked American: The Unexpected Decline of Leisure* London and New York: Harper Collins.
Shah, Hetan (2005) 'The politics of well-being' *Soundings*, 30, pp. 33–44.
Soper, Kate (1991) 'Greening prometheus', in Peter Osborne (ed.) *Socialism and the Limits of Liberalism* London: Verso, pp. 271–293.
Soper, Kate and Trentmann, Frank (eds) (2007) *Citizenship and Consumption* London: Palgrave.
Sustainable Development Commission UK (2003) 'Redefining Prosperity, Resource Productivity, Economic Growth and Sustainable Development' (www.sd-commission.org.uk/publications).
Thomas, Lyn (forthcoming, 2008) 'Alternative realities: Downshifting narratives in contemporary lifestyle television' *Cultural Studies*.
Trentmann, Frank (ed.) (2006) *The Making of the Consumer: Knowledge, Power and Identity in the Modern World* Berg: Oxford and New York.
Warde, Alan (1997) *Consumption, Food and Taste* London: Sage.
Warde, Alan (2004) 'Practice and field: Revising Bourdieusian Concepts' ESRC, CRIC Publication, April.
Williams, Raymond (1983) *Towards 2000* London: Chatto and Windus.
Williams, Raymond (1993; first published 1973) *The Country and the City* London: Hogarth.

Part I

Texts and Representations

1
Representing Consumers: Contesting Claims and Agendas

Roberta Sassatelli

In his widely known work *La société de consommation*, the French cultural critic Jean Baudrillard thus summed up contemporary society: 'Just as medieval society was balanced on God and the Devil, so ours is balanced on consumption and its denunciation' (Baudrillard, 1998, p. 196). Indeed, analysing a variety of discourses, we are starting to chart the cultural and political dynamics which have transformed the 'consumer' into a compelling cultural code for a variety of social institutions to mobilise people's desires and govern their practices, and indeed for people themselves to understand and manage their aspirations (Cohen, 2003; Daunton and Hilton, 2001; Sassatelli, 2007; Strasser *et al.*, 1998; Trentmann, 2005). The pro-consumerist rhetoric which emerged in the eighteenth century to justify market societies epitomised the entrenchment of a number of spaces of consumption (the coffee shop, the eating house, and so on) as part of a new sphere of legitimate action. Consumption was defined as a private matter, constructed as opposed to production, and envisaged as the pursuit of private happiness, indirectly, but firmly, linked to virtuous mechanisms in the public sphere. Within this framework, 'consumers' were constructed as private economic hedonists, preoccupied with individual pleasures and contributing both to the common good and to their own good, provided they behaved in disciplined ways within the rules of the market. The notion of the consumer, contested as it has always been, itself became an important device for social and cultural change.

Especially from the late nineteenth century onwards, a number of economic, cultural, and political agencies increasingly claimed for themselves the right and duty to address consumers and to speak for them. Advertising and marketing as well as state welfare agencies, consumer defence organisations, women's groups, consumer boycotts, and more recently the European Union, environmental groups, and new global movements have all contributed, together with social scientific discourses, to situate the 'consumer' as a fundamental subject-category within public discourse. Yet consumption has usually been envisioned through extreme rhetorical tropes. Different discourses and institutions have called forth different, often conflicting, images of the consumer – defining him/her as a sovereign of the market or a slave of commodities, a snob or a *flâneur*, a rebel or an imitator, a private collector or an entrepreneur with a sense of social duty, and encoding these images as belonging to a separate economic sphere or as moral and political figures in their own right.

In this light, this chapter looks at the cultural representation of consumption and the consumer. First, it considers the opposing discourses, both pro-consumerist and anti-consumerist, which have usually been deployed to portray consumption. Secondly, it looks at the swing of the pendulum between pro- and anti- consumerism, concluding that it has a generative role within contemporary culture. In particular, commercial advertising today appears to accommodate a plurality of images of what consumption is and does to people and the world. The new spirit of capitalism is increasingly attuned to a cultural critique of consumerism, and commercial images come to reflect the contested nature of commercialisation. Thirdly, this chapter concentrates on the critical framing of the consumer, as this is promoted by phenomena as diverse as Fair Trade, counter-advertising, alternative food networks, and so on. These developments appear to embrace new visions of the consumer that may represent a challenge to more established notions of market choice. They signal that the symbolic boundaries that have come to define the consumer as a specific economic identity, who lives in a private world removed from production and public concerns, are being destabilised. Overall, the chapter aims to discuss both how the consumer is a key social persona for contemporary culture and how there are many different visions of what the consumer is and ought to be. It aims to problematise Baudrillard's view that 'counter-discourse' does not afford 'any real distance' from (a single vision of) consumer society. While there may be no escape from market society and consumer choice, choices can be constructed and practised in quite a variety of ways, some of which seem to internalise values other than money and quantity and consider the common good, gift relations, and civic engagement as irreducible elements of consumers' gratification.

The pendulum of consumption

Generally speaking, in contemporary Western societies the 'consumer' has been regarded as playing a positive political-economic role: defined as he who buys goods for his personal use, he is often portrayed as the last resort to keep the economy turning whenever demand slackens. That commerce and consumption are the 'wheels of the market' is an idea that extends back to the origins of what we conventionally call 'modernity': attempting to account for and legitimate the new capitalist and bourgeois lifestyles, liberal theories have often taken on a genuine *pro-consumerist* character (Boltanski and Thévenot, 1991; Hirschman, 1977, 1982; Pocock, 1985). Apologists of the free market have typically claimed that consumerism is to be seen as a civilising force which pacifies societies. Consumption was in fact defined by its first apologists, much as it is by dominant contemporary economic theory, as the active search for personal gratification through commodities, and growth in personal consumption is seen as dangerous for neither the nation nor the individual (Appleby, 1993). Since humans are essentially defined as rational animals with infinite and undefined desires who have been able to guide the economy of nations to unimagined levels of prosperity, great care must be taken in assuring that this gratification is authentic. This generates a pressure to stress consumers' autonomy which culminates in the notion of 'consumer sovereignty'. Provided consumers are autonomous self-disciplined beings (something which is typically measured against their capacity to be model employees), their pursuit of happiness through a growth of private consumption is beneficial both to them and to the wider society. The idea of the sovereignty of the consumer has found numerous supporters well outside the science of economics: in politics, for example, as the citizen-consumer has been marshalled to reform welfare provision in the United Kingdom (Clarke, 2006; and more generally, see also Cohen, 2003). A pro-consumerist rhetoric is to be found above all in marketing and commercial advertising, which have had an important role in promoting consumption as an arena of legitimate action, rich in meaning and full of untarnished promises: a series of positive individual aspirations and images of self-realisation has been associated with the private acquisition and use of goods and services – happiness, sociability, youthfulness, enjoyment, friendship, eroticism, and so on.

Turning these views upside down, voices from a range of quarters have rallied to stigmatise consumption, casting it as a source of moral disorder, a soul-corrupting mirage. Indeed, whenever economic growth has opened up the availability of new consumer goods to upwardly mobile social groups or has threatened the traditional gender order, strong hostile sentiments towards material riches have emerged. These sentiments may well have a disciplining function if it is true, as Veblen (1994, p. 53) insisted, that 'consumption of luxuries in the true sense is a consumption directed to the comfort of the consumer itself and it is therefore a mark of the master': when practised by subordinated groups, luxury consumption tends to elicit moral suspicion and social control. It is therefore not surprising that consumption has often been seen as a negative expression of the triumph of the modern market which weakens men, turning them into useless citizens incapable of defending their own country or participating in politics, whilst making women superficial and idle, unfit as wives and mothers (Hilton, 2002; Hirschman, 1977, 1982; Horowitz, 1985; Sassatelli, 1997; Searle, 1998). According to an *anti-consumerist* rhetoric, consumption, deprecated in its modern guise as 'consumerism' or 'consumer culture', gave birth to spiritual impoverishment for which people sought comfort in material goods, a surrogate for traditional forms of satisfaction, self-realisation, and identification through work and political participation. The huge growth in material culture is thus criticised as a source of disorientation and a threat to the authenticity of the self, who should be strong and autonomous, able to become himself/herself through deeds and relationships and not through possessions. Within the Marxist tradition this growth has been described as a process of 'reification' in which human beings become quantifiable and fungible objects like commodities (Lukács, 1971). Consumerism is seen as promoting a 'narcissist personality type' (Lasch, 1991). Where the ascetic culture of production favours the development of strong personalities attached to duties and to the family, consumer culture favours the development of weak and isolated personalities, who continually search for gratification in objects and who are fated to be continually deluded: their desperate search for pleasure to fill their emptiness is in fact a form of 'aggression' which reduces everything to a commodity, an object interchangeable with other objects. The identification of a schizoid splitting of culture within modern capitalism and the repulsion felt for the voracity of the consumer are recurrent themes in academic literature and public discourse alike (Schor, 2007). They have often been articulated with a critique of the processes of bureaucratisation, rationalisation, and standardisation: the consumer has been depicted as someone who undergoes a senseless work routine only to get the money necessary to acquire more. The contemporary self is often portrayed as being constructed around contradictory demands which produce a 'bulimic personality type' imprisoned in a perverse circle of consumption and production (Bordo, 1993). In a similar vein, most authors of a critical persuasion regard advertising as the motor of this cycle: advertising is the ideological engine of a system in which work has lost its meaning, to which people nevertheless remain attached because they cannot give up the dreams associated with advertised goods.

Both pro- and anti-consumerist views provide a caricature of consumption, its practices and its meanings in everyday life. Apocalyptic views remove consumption from the web of social relations to criticise its social impact. The characterisation of consumer culture as totally opposed to work, as ruled by a consuming passion from which all inclinations towards moderation and rationality have been excluded, reduces people to mere viewers of adverts and is blind to the interchanges between goods and people: consumption often has a relational character (even if within rather narrow circles), and its pleasures are often related to gift relations or to aesthetic sensibilities that engage people's capacity to attribute value to the world (Douglas and Isherwood, 1979; Miller, 1987). On the other side of the fence, apologetic views of consumption are surprisingly blind to the social limits of consumption, including the fact that an increased expenditure does not necessarily lead to well-being and happiness. They divert attention from the fact that, for all the claims of liberal and neo-liberal slogans, the common good does not automatically spring forth from the pursuit of individual interests. Such a celebratory stance ignores our awareness that collective goods can rarely be provided efficiently by the market alone and that the democratisation of luxuries may become an environmentally dangerous and socially pointless game (Castells, 1977; Hirsch, 1977; Sen, 1985).

Broadly speaking, all of the discourses on consumption, even those providing a considered scholarly view, run the risk of falling into celebration or censorship, yielding to a series of binary oppositions which trap us in dichotomous thought: public/private, duty/pleasure, rational/irrational, and of course freedom/oppression. Both apocalyptic and apologetic views may be seen as opposed but complementary ways of consoling consumers, letting them catch a glimpse of worlds where absolute liberty exists and can be delivered through prescriptions as apparently simple as total rejection or total acceptance. This produces an ongoing evaluative pendulum, which is often exploited by commercial advertising to frame consumption. As cultural analysts, we are called on to consider the workings of this evaluative pendulum, but must avoid succumbing to it and to its tendency to strait-jacket discussion in terms of extreme emotional reactions.

Indeed, both pro-consumerist and anti-consumerist views take seriously the idea that consumer culture produces consumers, but their polemical view of consumer subjectivity does not account for the plurality and contested nature of consumption and for the irreducibility of consumers' practices to commercial representations. There is always a gap between the representation (in commercial or public discourse) of consumption and actual consumption practices, which are generated, even unwittingly, through everyday creativity (Willis, 1990). Rather than being based on a sovereign consumer or producing a pathological consumer as its norm, consumer culture offers visions of normality, happiness, and fairness which people are asked to engage with; and it accommodates a variety of cultures of consumption, which must enrich values other than mere affluence (Sassatelli, 2007). In doing so, these cultures draw on both pro- and anti-consumerist themes and help develop new rhetorical and conceptual tools for the framing of the consumer.

Productive contradictions

As suggested, the swing of the pendulum between apology and criticism of consumption appears to have a productive role within contemporary culture. The effects of this can be seen in commercial advertising and marketing. By and large, early marketing had its roots in Freudian psychoanalysis and in its attention to individual impulsions towards material pleasures (Bowlby, 1993). More prosaically, when marketing was born in the United States at the beginning of the twentieth century, the model of the consumer was that of someone trying to keep up with the neighbours: in order to create demand, advertising made the most of the feelings of inadequacy that characterised most Americans, people who were often social climbers, uprooted from their culture of origin and recently urbanised (Ewen, 1976). The emphasis on conformity and belonging was justified at the time by the necessity of creating a population of consumers corresponding to mass production, flattening the cultural heterogeneity typical of a country founded on the continuous immigration of different populations. The adverts of the time were thus both heralds and examples of that 'American Way of Life' which created national homogeneity above all through a style of consumption. Today, advertising rarely promulgates an idea of conformity. Indeed, as Frank has suggested in his discussion of the links between the advertising industry and 1960s counter-culture in the United States, fantasy and creativity, rebellion and non-conformity have become essential elements of the American culture of advertising production (Frank, 1997). Difference and eccentricity feature prominently in contemporary ads all around the world; anti-consumerist themes have also been digested by advertising and, as reality TV thrives, there seems to be a widespread attention to images drawn from 'real life'. These advertising modes of address call forth and respond to new sensibilities among consumers, in particular the cultured middle classes who have become more aware of the power of the media. In the late 1990s, the Body Shop gained international attention for the campaign 'Love your body', which featured a voluptuous plastic doll named Ruby. The print ads and posters showed Ruby reclining on a sofa under the headline: 'There are 3 billion women who don't look like supermodels and only 8 who do.' This initiated a 'real beauty' trend which is now being deployed in mainstream cosmetic advertising (for example, the Dove personal-care products sold by Unilever now claim to present women in advertisements as they are, rather than as some believe they ought to be), and which is getting institutionalised within the advertising profession itself: for some years now, the association Advertising Women of New York has presented awards to campaigns that portray women in non-stereotypical ways.

The original Body Shop advertisements counted on factors dissonant with the meanings that normally characterised cosmetics ads, and they captured the feelings of those women who, while being socially aware, critical and perhaps feminist, did not want to give up taking care of themselves. They worked both by positioning themselves against mainstream pro-consumerist advertising and by being part of it. In its turn towards anti-consumerist themes, it put consumers' increased social awareness to work to sell its products. Such an example highlights the complex cultural circuit in which advertising, advertising professionals, consumers, and everyday culture are interlocked. Indeed, it may be suggested that the emphasis on the active and critical consumer paramount in contemporary academic theories of consumption partly reflects and is generated by new consumer identities. The latter both emerge from grass-root social processes and are mobilised by the need of 'creative' advertising for equally 'creative' consumers. The vision of creativity promoted by the industry is seen as 'inclusivist' rather than 'exclusivist'; it broadens the meaning of 'creativity' to claim it for itself as well as for a larger set of activities, which include consumption and consumers' reading of ads (Nixon, 2003).

Arguably, Western middle-class consumers in particular are increasingly aware of the vulnerability of mainstream advertising. Contemporary advertisements' growing resort to indirect ways of conveying a message – such as humour, irony and paradox – may be considered evidence of this. Certainly, even consumers' capacity to appreciate more complex and engaged messages may be colonised (so to speak) by advertising. The development of 'postmodern' image-led advertising, where the products disappear behind the image (as in many of Benetton's campaigns), has been seen as an acknowledgment of consumers' interpretive powers and an attempt to put these powers at work for commercial ends (Morris, 2005). However, the critique of such campaigns as opportunist forms of commoditisation should be set against a wider picture. A study of the US discourse on the Benetton 'We on death row' campaign in the so-called 'prestigious press' shows that this campaign was a site where ideological differences between United States and Europe were played out (Kraidy and Goeddertz, 2003). The political potential of such advertisements is to be seen in the fact that the US press eagerly deployed 'commodification' as a powerful 'frame of othering' in order to portray Benetton as the foreign 'other' and push the issue of capital punishment which was bought up by the campaign to the margin of public discourse. Such an example helps us open up the complex cultural circuit of the commodity form. On the one hand, commercial images carry meanings and symbols which go well beyond their commercial aims. These meanings can be appropriated in quite subversive ways by consumers. On the other hand, consumers and producers can use commodities as a means to express political views or with clear political intentions. This in turn does not guarantee that the overall economic and political effects will be the desired ones.

Commodities and commoditisation are in this sense a rather 'indifferent' vector, whose meanings and significance in everyday life (the micro level) largely depend on contexts of consumption, and whose effects on the environment, the economy, or politics (the macro level) largely depend on how the whole commodity chain is organised. As much as they carry different meanings, commodities are never value-free, and cultural, economic, and political entrepreneurs have a prime role in pinning down their significance, be it political or otherwise. For example, in the Sicilian city of Palermo many stalls currently display a variety of T-shirts sporting references to the Mafia, romanticising it as a regional product and as a curiosity for tourists (for example, 'Mafia made in Italy'). These T-shirts have raised concerns among the public, in particular magistrates and public officials involved in the fight against the Mafia. One can easily discern the duplicity of the symbolic world which is brought to life by such a commercial operation: criminal associations can be exorcised but also legitimated, their symbols read with voyeuristic curiosity, and then either repudiated in life or experienced as a nostalgic phenomenon of the past or as an innocuous fact of the present. This however should not rule out the possibility that the market and the fight against the Mafia are intertwined in rather different ways. While the meanings of consumption are to a degree always open-ended and indeterminate, commodity circuits can be organised as to make social justice part of the economic calculus. Operating from Sicily but reaching the whole of the Italian market, the association 'Terra Libera!' offers a host of 'ethical' commodities (olive oil, pasta, beans, and so on), which are produced by social cooperatives managing land which has been confiscated from the Mafia. 'Terra Libera!' operates on solidarity and environmental principles on the production side, asks customers to purchase its products 'because it is possible to beat the Mafia in this way, too', and deploys sophisticated advertising techniques reaching a large public thanks to supermarket distribution. The symbolic campaigns against the Mafia which have been sponsored by this cooperative include 'Addiopizzo!' ('Goodbye, pizzo!' – 'pizzo' is the protection money traders pay to the Mafia). This uses – precisely – a T-shirt, incidentally one of the most banal items of mass consumerism, as a manifesto for critical consumption and a means of self-financing. Clearly, the simultaneous presence on the market of quite different T-shirts which work on the Mafia theme can make things more difficult for both concerned consumers and producers, yet in the 'Addiopizzo!' campaign consumerism appears to be harnessed to social justice; and justice is predicated on commercialisation, in the very attempt to tease out its political potential.

No easy road to Altermarket

Apocalyptic views tend to deny that the sphere of consumption can also constitute itself as a space for (different) forms of political action. Of course, commercial advertising teaches us that through commodity consumption (often an individualised or private activity) we can solve all problems – even social ones. In this sense, it does not favour traditional forms of political mobilisation (linked to the party, the workplace, or street protests). Consumption, however, is by no means just the exercise of private egotism. The political investment of consumption is something that a growing body of research on ethical and political purchases is documenting internationally (Ceccarini and Forno, 2006; Chessel and Cochoy, 2004; Micheletti *et al.*, 2004; Sassatelli, 2004, 2006a and b; Tosi, 2006). According to Kier, the sales volume of Fair Trade products grew 154% in Europe between 1997 and 2004 (Kier, 2005). Fair Trade coffee is the fastest growing segment in the US market, growing a spectacular 67% per year (Arnould, 2007). The European Social Survey has shown that approximately one-third of Europeans have boycotted certain goods or/and have bought goods for political and ethical reasons. According to Ifoam, organic production is growing a steady 10% every year and the growing number of studies on alternative food networks (from box schemes to farmers' markets) are showing their vigour in many advanced economies (Dubouisson-Quellier and Lamine, 2004; Goodman, 2003; Holloway *et al.*, 2006). While historically consumers have organised cooperatives to safeguard their purchasing power (Furlough and Strikwerda, 1999), and social movements have launched boycotts and have called forth the consumer as a political actor (Cohen, 2003; Friedman, 1999; Trentmann, 2005), the 1999 meeting of the World Trade Organisation opened a new phase, in which new social movements of an alternative-global variety have massively resorted to the whole spectrum of consumer actions (boycotts, naming and blaming, ethical merchandising, and so on) to widen the repertoire of political participation, find new ways to mobilise people, and address global issues.

A variety of actors (both individual and collective, economic and political, oriented towards profit maximisation or towards collective goods) are contributing to shaping alternative views of the market. This variety is reflected in the many nuances of the discourses on the 'critical' consumer, their uneven resonance, and the varying economic and political effectiveness of attempts to approach commodities as bearers of environmental, ethical, and political concerns. We can, however, identify a fundamental cultural theme at the outset, in that consumer choice is portrayed as neither universally good nor a private issue. Most forms of critical consumption share some kind of interest in environmental values and address both redistribution issues and the problems generated by the increased separation and disentanglement of production and consumption. People are typically asked to consume better. As a source of power, consumption is not to be given up altogether, but consumer choice is framed as a practice with direct and momentous consequences, capable of expressing consumer sovereignty only if consumers take full responsibility for the environmental, social, and political effects of their choices and are ready to reconsider their consuming life on those grounds.

Extensive empirical research conducted in Italy among engaged consumers as well as activists from a variety of organisations makes it clear that different initiatives share a distinct notion of consumer sovereignty which critically elaborates, and sometimes overturns, laissez-faire wisdom (Leonini and Sassatelli, 2008; Sassatelli, 2006a). Three themes in particular seem to emerge in varying degrees and combinations: redistribution and interdependency, collective goods, and the pleasures of frugality. Most informants put forward a civic vision of the market, contending that market relations thrive among equals, and indeed that to realise itself the market's social potential requires a pacified social space, which places value on redistribution and sees the powerful consumer as the prime motor behind this. They also share the view that goods which transcend individual, exclusive enjoyment (in particular, the environment) are of the essence for consumers' quality of life, but are all too often neglected by capitalist market relations: here again, consumer choice is seen as a way to internalise environmental factors. Finally, the liberal view of the relationship between consumption and happiness is regarded as simplistic.

This matches a growing body of literature in philosophy and the social sciences which argues that people's well-being might be understood in terms other than their expenditure, and which starts from notions of 'quality of life' which will often add environmental or communitarian depth to a short-term, individualist, and private vision of individual choice (Nussbaum and Sen, 1993). This may even imply some form of 'voluntary simplicity', 'sobriety', or 'downshifting' in consumption, rejecting upscale spending and long working hours, and living a simpler, more relaxed life in order to discover new pleasures and enhance personal satisfaction as well as to further socio-economic equality and environmental awareness (see also Etzioni, 1997; Nelson *et al.*, 2007; Soper, 2008).

The discourses surrounding critical consumer practices provide a set of specific criteria of choice drawing on 'regimes of justifications' (Boltanski and Thévenot, 1991) which have been taken beyond the dominant mode of legitimating markets in Western culture. As I have suggested elsewhere (Sassatelli, 2006b), themes mainly associated with the promotion of consumption as a legitimate sphere of action *per se* – 'taste', 'good taste', 'pleasure', 'fantasy', 'comfort', 'distinction', 'happiness', 'refinement', and so on – are replaced by themes predominantly associated with the definition of a democratic public sphere and with production. The vocabulary of critical consumerism draws either on social and political activism (to purchase is to 'vote', 'protest', 'make oneself heard', 'change the world', 'help the community', 'mobilise for a better future', and so on) or on production (to purchase here becomes 'work you do for the community', 'effort done for yourself and the other', 'creative', 'productive', and so on). Alongside this, a new set of criteria for measuring quality of life and pleasure is slowly evolving which draws on spiritual themes, de-rationalisation, and communitarianism.

While 'lifestyle politics' has come to function as a form of civic participation for many people (Beck and Gernsheim, 2001; Schudson, 1998), consumption has become so 'politicised' that it is no longer possible to sharply divide between 'citizenship and civic duty', on the one hand, and 'consumption and self-interest', on the other (Scammell, 2000; Soper and Trentmann, 2007). A number of voices have thus celebrated the political persona of the consumer. The consumer has been portrayed as the truly global actor who can be a counterweight to big transnational corporations and can replace the vanishing citizen in working for a cosmopolitan democracy: today, 'citizens discover the act of shopping as one in which they can always cast their ballot – on a world scale, no less' (Beck and Gernsheim, 2001, p. 44). Similar claims rests on a dubious metaphor assimilating consumer choice with voting (and voting with democracy). Let us tease out what is implicit in the metaphor: that consumer choice is like voting, that it is effective, that it is democratic. These assumptions can easily obliterate the specificity of both voting and consumer choice, the fruitful synergies which can be produced by considering their mutual transformation, and, last but not least, their ambiguities.

To be sure, it would be mistaken systematically to attribute a deliberately political intention to all consumer choices of a critical variety. Many of the practices which come under the umbrella of critical consumerism may be conducted by consumers who have in mind meanings and objectives other than strictly political ones. For example, alternative distribution networks, including second-hand shops, not only respond to a politically conscious middle-class consumer, but also attract disadvantaged urban groups who may not be able to afford to shop via formal channels (Williams and Paddock, 2003). Likewise, the demand for organically grown vegetables typically mixes private health concerns with some degree of environmental consciousness and comes from diverse sources, including a large vegetarian movement as well as health-conscious or gourmet carnivores (Lockie and Kristen, 2002). In Italy, a large proportion of those who buy Fair Trade goods in supermarkets (for example) do so because they 'like' the products or consider them 'better quality', or just 'by chance' (Leonini and Sassatelli, 2008). Still, alternative ways of consuming are not a syndrome, an ideology as Baudrillard maintained. Nor are they yet another positional option, the consumption of lost simplicity on luxury grounds. However, we should neither leave out of account the intrinsic pleasures of ethical shopping nor just equate 'genuine' critical choices with political votes.

Shopping ethically enables us to make choices which matter to us in ways that political voting may not, because these choices matter in themselves, empowering us in everyday life, rather than for their expressive potential or possible larger effects on macro-realities (Schudson, 2007). Indeed, peoples' lives can be entirely reorganised starting from a number of apparently banal choices, which often require and bring about a different management of time, of space, and of social relations (Leonini and Sassatelli, 2008). While certain consumption such as the consumption of news seems to be crucial for both civic engagement and truly political consumerism (Friedland *et al.*, 2007; Shah *et al.*, 2007), the presence of politically and ethically motivated cultural intermediaries and social movements appears crucial if the micro-politics of everyday life are to be translated into political pressure as such.

This brings us to the question of effectiveness, conceived in terms of public resonance, corporate change, and ultimately political-economic change. We know that as Fair Trade goes mainstream, it has had its difficulties in always keeping its promises to help producers in developing countries. Recent work on global anti-sweatshop campaigns (Micheletti and Stolle, 2007) and on their appropriation by US company American Apparel (Littler and Moor, 2008) seems to point to the fact that wide public resonance, and even commercial success, may not always correspond to a real improvement in the working life of garment workers. Alternative consumer practices can easily be absorbed by the market. The marketing and advertising industries are well aware of the interest in ecological, ethical, and political themes among a certain strata of Western populations and have long started to promote their own versions of the 'greening of demand' (Zinkhan and Carlson, 1995). The institutionalisation of a dialogue between consumerist and environmental organisations and large multinational commercial companies may also have ambiguous effects (Barnett and Cloke, 2005; Doubleday, 2004). Codes for ethical business and for socially responsible management are becoming widespread, yet they are typically self-administered by industry itself. In response to boycotts and consumer choices in pursuit of specific causes, a variety of labelling schemes, often set up by ad hoc organisations variously linked with either business or political institutions, are playing a crucial role. This does not mean that ethical claims can easily be used in a purely instrumental fashion, for ethically oriented consumers may demand proof of standards and may push companies much further than expected. But it does suggest that it is unrealistic to imagine that there is a direct and symmetrical demand/supply relation between consumers and producers. In particular, the reaching of global markets may imply an emphasis on efficiency and promotion which can transform green and Fair Trade products into fetishes (Hudson and Hudson, 2003; Levi and Linton, 2003).

At the same time, the antinomy between commercial aims and ethical aims may become a dialectical resource to modify views of capitalism. This is certainly the fundamental dynamic of value creation and progress within the Fair Trade field (Leonini and Sassatelli, 2008). Italian activists have stressed the importance of keeping these values in synergy if Fair Trade is to be made viable and meaningful; they have used this antinomy as the basis for distinguishing between 'real' critical consumers ('activists' and those who are 'committed') and 'lifestyle' or 'fashion-oriented' consumers who are ready to jump on the bandwagon of Fair Trade. The different actors who occupy different positions in the field of alternative and critical consumption have expressed this dichotomy in various ways. Activists who work at the commercial end of Fair Trade (shops, import organisations) place emphasis on the positive role of commercialisation as 'cultural vector'; those who are concerned with labelling schemes stress the role of 'good principles'; and the cultural and political entrepreneurs emphasise the risks of commercialisation and the role of 'education and awareness'. The very fragmentation of the Fair Trade market in Italy seems to favour such a plurality of voices, which arguably results in a more democratic space. As has been documented, markets are indeed institutions that can be organised differently (see Callon, 1998; Carrier, 1997), and they can thus be put to many different ends. While there is no easy road to Altermarket, capitalist, profit-driven markets can be transformed to take into account the redistribution of resources, avoid economic polarisation, and stress a new set of pleasures. Engaged consumers may be one of the levers of the transformation.

Concluding reflections

Finally, let us consider the relationship between critical consumption and democracy. Apocalyptic views of consumerism are largely one-sided in excluding any possible contribution of consumption to civic life. Yet it is likewise mistaken to suppose that the individual, global consumer increasingly invoked by Fair Trade and other critical initiatives can now carry on a global scale the duties and capacities of the citizen or can transform the awareness of effects into a politics of justice. To be sure, Eric Arnould has a point when he maintains that the 'successful progressive practice of citizenship should take place through market mediated forms in our culture because these are the templates for action and understanding available to most people' (p. 106). Still, while we all consume, we do so in many different ways which are largely a function of our different resources and of the different political infrastructure that upholds different systems of consumption. In very banal terms, if in contemporary democracies each citizen has a vote, consumers are notably different in terms of purchasing power and may thus have rather different degrees of influence on the market. Critical consumption does try to make the political infrastructure of our everyday consumer lives visible to us, yet different positions within that infrastructure, and different infrastructures (often still determined by national borders), make critical consumption more or less probable and viable. This double-bind construction shows that there is still a space for politics in the traditional sense; indeed, that politics and consumption can act in synergy in the transformation of the market.

Acknowledgements

I would like to thank all participants at the workshop on critical consumption which I organised in February 2007 at Birkbeck College during my ESRC/AHRB International Fellowship within the 'Cultures of Consumption' Programme. A particular thanks is owed to Stephanie Nixon for her assistance in the organisation of the seminar, as well to and Clive Barnett and Frank Trentmann for their suggestions.

References

Appleby, J. O. (1993) 'Consumption in early modern social thought', in J. Brewer and R. Porter (eds) *Consumption and the World of Goods*, London: Routledge.

Arnould, E. J. (2007) 'Should Consumer Citizens Escape the Market?' *Annals AAPSS*, 611, May, pp. 96–111.

Barnett, C. and Cloke, P. (2005) 'Consuming ethics: Articulating the subjects and spaces of ethical consumption' *Antipode*, 27(1), pp. 23–45.

Baudrillard, J. (1998; first published 1970) *The Consumer Society: Myths and Structures* London: Sage.

Beck, U. and Gernsheim, E. (2001) *Individualisation* London: Sage.

Boltanski, L. and Chiapello, E. (1999) *Le nouvel esprit du capitalisme* Paris: Gallimard.

Boltanski, L. and Thévenot, L. (1991) *De la justification. Les économies de la grandeur* Paris: Gallimard.

Bordo, S. (1993) *Unbearable Weight. Feminism, Western Culture and the Body* Berkeley: University of California Press.
Bowlby, R. (1993) *Shopping with Freud* London: Routledge.
Callon, M. (ed.) (1998) *The Laws of the Market* Oxford: Blackwell.
Carrier, J. (ed.) (1997) *Meanings of the Market* Oxford: Berg.
Castells, M. (1977; first published 1972) *The Urban Question* London: Edward Arnold.
Ceccarini, L. and Forno, F. (2006) 'From the street to the shops: The rising of new forms of political action in Italy' *South European Society and Politics*, 11(2), pp. 197–222.
Chessel, M-E. and Cochoy, F. (2004) 'Autour de la consommation engagé' *Sciences de la Société*, 62, pp. 3–14.
Clarke, J. (2006) 'Consumers, clients or citizens? Politics, policy and practice in the reform of social care' *European Societies*, 8(3), pp. 423–442.
Cohen, E. (2003) *A Consumers' Republic. The Politics of Mass Consumption in Postwar America* New York: Knopf.
Daunton, M. and Hilton, M. (eds) (2001) *The Politics of Consumption* Oxford: Berg.
Doubleday, R. (2004) 'Institutionalizing non-governmental organization dialogue at Unilever: Framing the public as "consumer-citizens"' *Science and Public Policy*, 31(2), pp. 117–126.
Douglas M. and Isherwood, B. (1979) *The World of Goods. Towards an Anthropology of Consumption* New York: Basic Books.
Dubouisson-Quellier, S. and Lamine, C. (2004) 'Faire le marché autrement. L'abonnement à un panier de fruit e de légumes comme forme d'engagement politique des consommateurs' *Sciences de la Société*, 62, pp. 145–168.
Etzioni, A. (1997) 'Voluntary simplicity: Characterization, select psychological implications, and societal consequences' *Journal of Economic Psychology*, 19(5), pp. 619–643.
Ewen, S. (1976) *Captains of Consciousness. Advertising and the Social Roots of Consumer Culture* New York: McGraw Hill.
Frank, T. (1997) *The Conquest of Cool. Business Culture, Counter Culture and the Rise of Hip Consumerism* Chicago: Chicago University Press.
Friedland, L., Shah, D. V., Lee, J-M., Rademacher, M. A., Atkinson, L. and Hove, T. (2007) 'Capital, consumption, communication and citizenship: The social positioning of taste and civic culture in the United States' *Annals AAPSS*, 611, pp. 31–50.
Friedman, M. (1999) *Consumer Boycotts* New York: Routledge.
Furlough, E. and Strikwerda, C. (1999) *Consumers Against Capitalism? Consumers' Cooperation in Europe, North America and Japan, 1840–1990* Oxford: Rowan and Littlefield.
Goodman, D. (2003) 'The quality turn and alternative food practices. Reflections and agenda' *Journal of Rural Studies*, 19, pp. 1–7.
Hilton, M. (2002) 'The female consumer and the politics of consumption in twentieth-century Britain' *The Historical Journal*, 45(1), pp. 103–128.
Hirsch, F. (1977) *The Social Limits of Growth* London: Routledge.
Hirschman, A. O. (1977) *The Passions and the Interests: Political Arguments for Capitalism before its Triumph* Princeton: Princeton University Press.
Hirschmann, A. O. (1982) 'Rival interpretations of market society: Civilizing, destructive or feeble' *Journal of Economic Literature*, 20, pp. 1463–1484.

Holloway, L., Cox, R., Venn, L., Kneafsey, M., Dowler, E., and Tuomainen, H. (2006) 'Managing sustainable farmed landscape through "alternative" food networks. A case study from Italy' *The Geographical Journal*, 172(3), pp. 219–229.

Horowitz, D. (1985) *The Morality of Spending* Baltimore: Johns Hopkins University Press.

Hudson, I. and Hudson, M. (2003) 'Removing the veil?' *Organization and Environment*, 16(4), pp. 423–430.

Kier, J. M. (2005) *Fair Trade in Europe 2005* Brussels: EU Trade Commission.

Kraidy, M. M. and Goeddertz, T. (2003) 'Transnational advertising and international relations. US press discourses on the Benetton "we on death row" campaign' *Media, Culture and Society*, 25, pp. 147–165.

Lasch, C. (1991; first published 1979) *The Culture of Narcissism* New York: Norton.

Leonini, L. and Sassatelli, R. (2008) *Cittadini & Consumatori* Rome: Laterza.

Levi, M. and Linton, A. (2003) 'Fair trade, a cup at a time?' *Politics & Society*, 31(3), pp. 407–432.

Littler, J. and Moor, L. (forthcoming in 2008) 'Fourth worlds and neo-Fordism: American apparel and the cultural economy of consumer anxiety' *Cultural Studies*, 22, 5–6.

Lockie, S. and Kristen, L. (2002) 'Eating green' *Sociologia Ruralis*, 42(1), pp. 23–40.

Lukács, G. (1971; first published 1923) *History and Class Consciousness: Studies in Marxist Dialectics* London: Merlin.

Micheletti, M., Follesdal, A., and Stolle, D. (eds) (2004) *Politics, Products and Markets* London: Transaction Publishers.

Micheletti, M. and Stolle, S. (2007) 'Mobilizing consumers to take responsibility for global social justice' *Annals AAPSS*, 611, pp. 157–175.

Miller, D. (1987) *Material Culture and Mass Consumption* Oxford: Basil Blackwell.

Morris, M. (2005) 'Interpretability and social power, or why postmodern advertising works' *Media, Culture and Society*, 27(5), pp. 697–718.

Nelson, M. R., Rademacher, M. A. and Hy-Jin Paek (2007) 'Downshifting consumers = upshifting citizens? An examination of local freecycle community' *Annals AAPSS*, 611, May, pp. 1–56.

Nixon, S. (2003) *Advertising Cultures* London: Sage.

Nussbaum, M. and Sen, A. (eds) (1993) *The Quality of Life* Oxford: Clarendon Press.

Pocock, J. G. A. (1985) *Virtue, Commerce, History* Cambridge: Cambridge University Press.

Sassatelli, R. (1997) 'Consuming ambivalence: Eighteenth century public discourse on consumption and Mandeville's legacy' *Journal of Material Culture*, 2(3), pp. 339–360.

Sassatelli, R. (2004) 'The political morality of food. Discourses, contestation and alternative consumption', in M. Harvey, A. McMeekin and A. Warde (eds) *Qualities of Food*, Manchester: Manchester University Press.

Sassatelli, R. (2006a) 'Alternativi e critici. Consumo consapevole e partecipazione politica', in E. Di Nallo and R. Paltrinieri (eds) *Cum Sumo. Prospettive di analisi del consumo nella società globale* Milan: Angeli, pp. 386–401.

Sassatelli, R. (2006b) 'Virtue, responsibility and consumer choice. Framing critical consumerism', in J. Brewer and F. Trentmann (eds) *Consuming Cultures, Global Perspectives* Oxford: Berg.

Sassatelli, R. (2007) *Consumer Culture. History, Theory, Politics* London: Sage.
Scammell, M. (2000) 'Internet and civic engagement: The Age of the Citizen Consumer' *Political Communication*, 17, pp. 351–355.
Schor, J. (2007) 'In defense of consumer critique: Revisiting the consumption debates of the twentieth Century' *Annals AAPSS*, 611, pp. 16–30.
Schudson, M. (1998) *The Good Citizen. A History of American Civic Life* New York: Free Press.
Schudson, M. (2007) 'Citizens, consumers and the good society' *Annals AAPSS*, 611, May, pp. 236–249.
Searle, G. (1998) *Morality and the Market in Victorian Britain* Oxford: Clarendon Press.
Sen, A. K. (1985) *Commodities and Capabilities* Amsterdam: Elsevier.
Shah, D. V., McLeod, D. M., Friedland, L. and Nelson, M. R. (2007) 'The politics of consumption/The consumption of politics' *Annals AAPSS*, 611, May, pp. 6–15.
Soper, K. (forthcoming in 2008) 'Alternative hedonism, cultural theory and the role of aesthetic revisioning' *Cultural Studies*, 22, 5–6.
Soper, K. and Trentmann, F. (2007) (eds) *Citizenship and Consumption* London: Palgrave.
Strasser, S. McGovern, C. and Judd, M. (eds) (1998) *Getting and Spending: European and American Consumer Societies in the Twentieth Century* Cambridge: Cambridge University Press.
Tosi. S. (ed.) (2006) *Consumi e partecipazione politica. Tra azione individuale e mobilitazione collettiva* Milano: FrancoAngeli.
Trentmann, F. (ed.) (2005) *The Making of the Consumer: Knowledge, Power and Identity in the Modern World* Oxford: Berg.
Veblen, T. (1994; first published 1889) *The Theory of the Leisure Class* London: Macmillan.
Williams, C. C. and Paddock, C. (2003) 'The meaning of alternative consumption practices' *Cities*, 20(5), pp. 311–319.
Willis, P. (1990) *Common Culture* Milton Keynes: Open University Press.
Zinkhan, M. G. and Carlson, L. (1995) 'Green advertising and the reluctant consumer' *Journal of Advertising*, 24(2), pp. 1–16.

2
The Past, the Future and the Golden Age: Some Contemporary Versions of Pastoral

Martin Ryle

Anyone who laments a life now lost to us, and claims that we lived more happily when we lived 'closer to nature', is likely to be reminded that such claims and laments have been heard for hundreds of years. Literary evocation of the Golden Age, a blissful era where good things were abundant and life was simple ('sustainable', as we now say), goes back to classical times (Lerner, 1972; Williams, 1993). Today, imagery of Arcadia and the golden age is recycled in TV and magazine adverts, as 'pastoral kitsch...for the promotion of "country" products' (Garrard, 2004, p. 48). Sophisticated readers, therefore, are on their guard when it comes to pictures of simple rural life, especially when seen in a backward-looking perspective. Critics of the present are cautious about claiming anything was better in the past.

Meanwhile, however, we are more aware than ever of ecological limits and the threats we face because we will not respect them. Most of today's environmental destruction is wrought by forms of production and consumption unknown before the industrial revolution – in many cases, unknown before the mid-twentieth century. Because citizens of wealthy nations cannot go on consuming as we are, we might look again for old pleasures to complement and replace some of those we have grown accustomed to during the last 60 years. We need to sharpen our awareness of forms and possibilities of pleasure which have lately been compromised, diminished or destroyed by what is called progress. It is in these terms that my argument is related to the overall concerns of this book.

Even the most compelling representations of the world we have lost may seem remote from the preoccupations of our everyday lives, and its sensuous and aesthetic pleasures may seem to be those of nostalgia and escapism. However, as Terry Gifford notes, pastoral does not just offer imaginary escape from present realities. Versions of pastoral can also 'imply a better future conceived in the language of the present', and this is the source of the genre's 'oppositional potential' (Gifford, 1999, pp. 22, 36). It is in this perspective that I here consider some contemporary meanings, and examples, of pastoral, as a mode of cultural critique that involves imaginative retrospection. In doing so, I hope to counter the rather dismissive attitudes that critics since the 1970s (for example, Robert Hewison (1987) and Martin Wiener (1981) in two influential books of the Thatcher decade) have often held towards 'retrospective radicalism'.

'Retrospective radicalism' is a term coined by Raymond Williams in *The Country and the City*, first published in 1973; and it is with his discussion that I begin. Having emphasised the political context of Williams' developing argument and argued that he came to be at least as critical of progress as of retrospection, I go on to suggest how the figure of 'the border' which is so important in his book may help us to represent our own historical moment. I then turn to pastoral and neo-pastoral elements in some recent European novels, discussing their evocation of worlds and pleasures that are lost, destroyed and pre-empted by the way we live now. The three texts on which I focus, by Kazuo Ishiguro, Michel Houellebecq, and Ali Smith (Houellebecq, 2003; Ishiguro, 2005; Smith, 2006), are quite diverse in tone, but each raises questions about nature, culture and the environment as part of a troubled reflection on the likely future of the overdeveloped metropolis.

'Retrospective radicalism'

Williams' discussion of pastoral and nature writing in *The Country and the City* illuminates the broad political implications of the topic. After more than three decades, it retains a critical edge and sophistication often lacking in later, nature-focused eco-criticism, especially from the United States (for a recent critique of this, see Kerridge, 2001). In historical and textual readings based largely on works in English since the late sixteenth century, Williams shows how literary 'idealisations' of rural life have been 'partial and misleading responses' to the transformation of urban–rural relations in the capitalist order (Williams, 1993, p. 37). However, he does not oppose 'idealisation' absolutely to a reality that writing never grasps. Literature may be nostalgic, it may serve the powers that be, but it can also bring critical understanding. As if to show country writing at work, Williams' first chapter evokes the Welsh borders where he grew up and the Cambridgeshire landscape that surrounds him now: 'the elms, the may, the white horse, in the field beyond the window where I am writing' (p. 3).

Williams declared this moment of the book's making (1973) to be a critical moment. In his fourth chapter, 'Golden Ages', he refers to 'an evident crisis of values in our own world', as 'the capitalist thrust' continues to subjugate both town and country to its 'crude moneyed order' (p. 35). He then asks whether the 'retrospective radicalism' of the ruralist tradition can provide a basis for resistance and dissent. He identifies a purely conservative ruralism, with fascistic 'blood and soil' tendencies: here, lament for the old natural order implies a pernicious call for the reinstatement of oppressive social relations. However, there is also 'a precarious but persistent rural-intellectual radicalism: genuinely and actively hostile to industrialism and capitalism; opposed to commercialism and to the exploitation of environment; attached to country ways and feelings'. Williams evidently feels some sympathy with this tradition, but questions its relevance. 'The moment comes when any critique of the present must choose its bearings, between past and future' – and 'rural-intellectual radicalism' has chosen the past (p. 36).

Williams was a socialist, deeply aware of how millions of working-class families like his own had made their way out of poverty thanks to the struggles and victories of the labour movement. He was not inclined to dismiss lightly the uses of economic growth and technological innovation. The argument in 'Golden Ages', as we have traced it so far, seems to be clearing the ground: having consigned ruralism to the past, we expect Williams to identify socialism as its forward-looking alternative. However, what follows is a critique of those socialists who – in an ambivalence traced back to the historical argument of Marx – praise the 'progressive' character of industrial capitalism even as they damn it for 'its long record of misery in factories and towns':

> We hear again and again this brisk, impatient and it is said realistic response: to the productive efficiency, the newly liberated forces, of the capitalist breakthrough... an unreflecting celebration of mastery – power, yield, production, man's mastery of nature – as if the exploitation of natural resources could be separated from the accompanying exploitation of men... [It is then argued that] at a certain stage... capitalism begins to lose this progressive character and for further productive efficiency, for the more telling mastery of nature, must be replaced, superseded, by socialism.
>
> (p. 37)

Directed explicitly at 'certain metropolitan intellectuals' (p. 36), Williams' critique was surely aimed also at the left-of-centre mainstream. The British Labour governments led by Harold Wilson and James

Callaghan in the late 1960s and the 1970s (whose 'socialism' in no way aimed to 'supersede' capitalism, but which had temporarily taken over its management) were entirely committed to 'productive efficiency' and 'mastery of nature'. In the decade after *The Country and the City*, Williams would go on to argue in directly political terms against this dominant strand of Labourism and would insist that – given 'the newly realized and decisive fact: that we cannot *materially* go on in the old ways' – a renewed left politics must place environment and ecology at the heart of its concerns (Williams, 1981, p. 148f; 1983: for the political context, see Ryle, 1988). The 'crisis' of the socialist tradition lay, then, in the fact that its supposedly progressive programmes only intensified, in 'more telling mastery', the old destructive production. What has been lost turns out to be of vital importance in our thought about what is to come: only if we acknowledge the record of destruction can we hope to make an alternative.

Williams' voice was as prescient as it was unfashionable. The conflict between productivist 'mastery' and ecological limits, which he noted in 1973 and insisted on through the 1980s until his death in 1988, is only now being registered – belatedly and feebly – in the formal mainstream politics of most over-developed societies (including Britain and the United States). The antinomies of destructive production have been repressed; but they are returning to haunt the ways we understand both progress and retrospect.

Borders in space and time

In *The Country and the City*, Williams concludes his critique of socialist complicity in the 'celebration of mastery' by acknowledging that 'against this powerful tendency... even the old, sad, retrospective radicalism seems to bear and to embody a human concern' (Williams, 1993, p. 37). But he declares that 'in the end it cannot do this' and turns, in the remainder of the book, away from the political context of its composition and towards his literary texts and the history they (mis)represent:

Without following the detail of Williams' readings, we can note that many writers of whom he speaks eloquently and warmly (including Thomas Hardy, D. H. Lawrence and the colonial and post-colonial novelists discussed in the penultimate chapter) are situated on 'borders' – places exposed to, but not yet wholly subdued by, the 'progressive thrust' (p. 37) of metropolitan capitalism. Nineteenth-century Dorset is an 'old rural world', but in Hardy's major novels, its people encounter 'experiences of change and of the difficulty of choice'. Lawrence 'lived on a border... In his own development... he was on a cultural border'. In Chinua Achebe's Nigerian trilogy, we 'see... from within a rural community as the white men – missionaries, district officers – arrive with their mercenary soldiers and police': these, writes Williams, are 'bitterly remembered experiences at the receiving end of the process' (pp. 197, 264, 285: and see McIlroy, 1993, p. 269).

Williams' 'border' is spatial-temporal, a place condemned and privileged to a particularly intense experience of historical time. Those who live there know – more intimately than the metropolitan agents, the 'soldiers and police', could understand – that 'the progressive thrust' destroys as well as creating and that what it relegates to the past is nothing less than a world. The writer straddles the border, has allegiances on both sides and speaks across it. Having learnt to use the language of the metropolitan centre, he or she is at once inside and outside the process that transforms the native place, as Gillian Beer (1988) emphasises in her discussion of Hardy's (1995) *The Return of the Native*. Careless readers in the city may have misheard Hardy as 'the last voice of an old rural civilisation', just as a recent reviewer mistakes the Irish writer John McGahern's border country of Roscommon and Leitrim for 'a primitive community locked in a timeless past' (Patterson, 2006; Williams, 1993, p. 197). To the contrary, what distinguishes the writer's borderland is by no means 'timelessness': rather, it is the interaction there of the past, the present and the future. This border sensibility, and the 'structure of feeling' it draws on and embodies, catches history in its making and unmaking (on 'structures of feeling', see Williams, 1977, p. 128ff). I find that writing informed by this sensibility can at its best offer a more compelling reflection of the present – of the past we are outstripping, the choices we confront and the future our choices portend – than is found in the genre of apocalyptic prophecy often used to represent ecological crisis (see the thoughtful, critical discussion of 'Apocalypse' in Garrard, 2004, pp. 85–107).

We may think that these kinds of borderland are no longer to be found in first-world societies, whose affluent citizens are literally and virtually hyper-mobile in their globalised oyster. As geographical distance loses much of its significance, education and self-advancement need no longer entail quitting a marginal life-world to form a new and often ambivalent allegiance to a metropolitan centre, in the emblematic journey recorded in many novels of development (for example, by Bronte, Byatt, Dickens, Emecheta, Gissing and Lawrence). To be more at home everywhere is to be less at home in, and less exiled from, any particular place.

However, if the border we live on has been losing its spatial coordinates, its temporal aspect is felt all the more powerfully. In earlier periods, self-exiled city-dwellers might revisit the rural birthplace and re-enter their biographical and historical past: the sensuous life enfolded there could not just be evoked, it could be experienced again. As the capitalist-metropolitan 'thrust' extends its scope and speeds its pace, landscapes that embody alternative life-worlds have grown fewer, more remote and more vulnerable. Nowadays, the native borderland may become all but unrecognisable in the course of a lifetime. McGahern begins his *Memoir* (2005) by reflecting on the survival of the Leitrim landscape: 'The very poorness of the soil saved these fields when old hedges and great trees were being levelled throughout Europe for factory farming, and, amazingly, amid unrelenting change, these fields have hardly changed at all since I ran and played and worked in them as a boy' (McGahern, 2005, p. 1). More usual is the experience of George Bowling in George Orwell's novel *Coming up for Air*, which came out on the eve of the Second World War. All through his dispiriting years in suburbia, Bowling has cherished the image of rural Lower Binfield where he grew up, but when he tries to go back there, it has disappeared, 'buried somewhere in the middle of a sea of bricks' (Orwell, 1971, p. 180). His friend Porteous, a retired English teacher, likes quoting Wordsworth; and although Bowling scoffs at romantic ideas about children (p. 75), the landscape of his own childhood stands, in the economy of the novel, for the remembered 'tree', the 'single field', of Wordsworth's 'Immortality Ode', which 'speak of something that is gone' (Wordsworth, 1965). But while Wordsworth's tree is a living emblem, in Orwell the landscape has vanished.

The borderland which once reached back into the past has been wholly taken over into the present and the future, and that, quite apart from the lapse of time which separates adult from child, is why its pleasures are lost. In literary tradition, the 'clouds of glory' and the 'visionary gleam' that surround the child at play evoke a primal lost world: the irrecoverable infancy of the self and of humanity. We see now that they have also evoked more ordinary enchantments: skating on a lake at night (in Wordsworth's *Prelude*), or taking the jennet and cart, in 'perfect weather for cutting and gathering sticks', over the frost-bound Leitrim fields where McGahern was growing up (McGahern, 2005, p. 98). What progress has destroyed is not Arcadia – that is always there, and always lost – but the world in which these pleasures, and others like them, lay at our doors. It is time to read the image of lost rural pleasure in a fully historical as well as a mythographic (or mytho-sceptical) spirit: as representing an order of sensuous experience dependent on a material reality that is vanishing – on empty roads, on silence and darkness, on wide spaces of ungoverned time, on a landscape not parcelled up for use and profit.

McGahern's *Memoir* and his last novel *That They May Face the Rising Sun* (2002) are fully centred in geographical borderlands. By contrast, the novels that I now discuss (like Orwell's) have their centre of gravity in the metropolis – Paris, London, the motorways of England. Here, marginal landscapes which evoke the past make important and ambiguous appearances. These are documents of contemporary 'border sensibility', in which the backward look plays a decisive part in grasping the historical present.

Three novels

My novel-texts are all recent: Michel Houellebecq's *Platform* (Paris, 2001; London, 2003), Kazuo Ishiguro's *Never Let Me Go* (2005), and Ali Smith's *The Accidental* (2005). All are of great interest in terms of my theme, but space allows me to discuss only selected aspects of each.

Houellebecq's fiction stages an ambiguous, equivocal challenge to the contemporary pleasure-loving self (Ryle, 2004). Michel Renault, in *Platform*, typifies the metropolitan subject of consumerist hedonism, and like Houellebecq's other first-person protagonists, he speaks a language tense with ideological abrasion, at once asserting, mocking, and subverting the 'values' by which he lives. A semi-dissenting specimen of the class that he embodies and describes, he is an agent of (albeit cynical) critique, and himself an (albeit self-aware) product of consumerism.

Platform raises the question of whether earlier ideas of the good life and forms of pleasure were less destructive than those which well-off Europeans pursue today. Is the present/future, whose equivocal celebration dominates the novel's foreground, altogether preferable to the past which appears in its margins? The question is brought to a focus in the central theme of tourism, including sex tourism. 'Unspoiled' third-world locations are physically and culturally remodelled, making their beaches, and the bodies of their young service workers, newly available for first-world visitors to enjoy.

During a package holiday in Thailand, Renault and his party spend a night in an 'ecological paradise', where (the guide assures them) they will go 'absolutely back to nature' (Houellebecq, 2003, p. 62). The hotel is 'superb': 'Small, beautifully sculpted chalets made of teak, connected by a pathway decked with flowers, overhung the river.' But Renault's summary of its history strikes a note of disdain:

> A brochure in my hotel room gave me some information about the history of the resort, which was the product of a wonderful human adventure, that of Bertrand Le Moal, backpacker *avant la lettre* who, having fallen in love with this place, had 'laid down his pack' here at the end of the 60s. With furious energy, and the help of his Karen friends, little by little he had built this 'ecological paradise', which an international clientele could now enjoy.
>
> (p. 66)

What tourists consume as authentic has been remade and marketed by first-world capital. Pastoral, grown exotic, is consumed at the far end of a jet flight. This applies with especial irony here, where the entrepreneur was a simple-lifer, from the hippy decade; but such versions of paradise are for sale in many resorts where the waves of tourism wash. It is not only faraway places that are changed by the air freighting of Europeans to the Oriental sun: in chapter one, Renault watches a TV nature film about 'silurids – huge fish with no scales which had become common in French rivers as a result of global warming' (p. 10).

This, global warming and all, is the present/future. The past is represented by two residual cultural zones within France: marginal places, a few score kilometres from Paris, but remoter than a Thai beach from the metropolitan pleasure-world, if we measure not in space but in time. Valérie, Renault's lover, was brought up in rural Brittany: the price of pork fell and her father sold his pig-farm, investing part of the proceeds in studio flats in Torremolinos, which brought in more money than he had ever made from farming (pp. 53–55). Two of Renault's fellow-tourists are from the borderland of the Jura: he refers to them as 'the ecologists... living in their godforsaken hole in the Franche-Comté' (p. 102).

Tourism – the world's biggest and fastest-growing 'industry', as Houellebecq reminds us (p. 29) – creates a paradoxical geography. European tourists travel thousands of miles to go 'absolutely back to nature' in newly built 'ecological paradises': the motif of sex tourism suggests that these affluent visitors embody a consumerism apt to transgress even the few ethical norms that used supposedly to govern it. Meanwhile, landscapes of Europe's rural past, disregarded, despised, and perhaps in some ways despicable, survive, or decay, much closer to home. Economics is a 'mystery', says Renault, that he has 'never really understood' (pp. 83, 214): but Houellebecq shows how this uneven, and destructive, pattern of development follows from the economics of tourism as now enjoyed by well-off global citizens.

Franche-Comté is as remote, and as 'backward', as McGahern's Leitrim; which consigns it, for Renault, to an irrelevant past. In *Never Let Me Go* and *The Accidental*, there is a similar spatialisation of time. Both books contrast the marginal landscape of Norfolk to familiar metropolitan spaces (Ishiguro's motorways and drab institutional buildings, Smith's North London of the comfortable professional classes). Both novels have a significant 'pastoral' aspect, in Gifford's broad sense: 'pastoral refers to any literature that describes the country with an implicit or explicit contrast to the urban' (Gifford, 1999, p. 2). In both, events in the margins trouble and disturb metropolitan norms.

The Norfolk of *Never Let me Go* is 'a peaceful corner... but also something of a lost corner' (Ishiguro, 2005, p. 60). The narrator, Kathy, travels there at the close of the book, to stand at the edge of a field and remember her dead friend and lover, Tommy:

> I found I was standing before acres of ploughed earth. There was a fence keeping me from stepping into the field, with two lines of barbed wire, and I could see how this fence and the cluster of three or four trees above me were the only things breaking the wind for miles. All along the fence, especially along the lower line of wire, all sorts of rubbish had caught and tangled. It was like the debris you get on a sea-shore: the wind must have carried some of it for miles and miles before finally coming up against these trees and these lines of wire. Up in the branches of the trees, too, I could see, flapping about, torn plastic sheeting and bits of old carrier bags.
>
> (p. 263)

This scene certainly lacks the charm of McGahern's lakes, skies and fields. We are not in Leitrim; we are on the borders of metropolitan England. Since leaving school and starting work, Kathy has spent much of her time driving from place to place on motorways and dual carriageways, often unable to stop and get out. Still, her preference for 'obscure back roads' marks her out: ' "Kath, you really know some weird roads" ', says Tommy (p. 249). What she seeks and finds in Norfolk is not an 'ecological paradise' – no such place exists in the novel's geography – but a field where the sea and the wind can still bring an intimation of mortality. A similar journey taken shortly before Tommy's death leads Kath, Tommy and Ruth, the main characters, through 'featureless countryside' on a 'near-empty road' to another liminal space, half-land and half-water, where they find a 'beached boat' (pp. 201, 205): a cold emblem of mortality and also, because it recalls the death-ships of ancient seafarers, a token of how human culture has sought to come to terms with death by acknowledging it.

Here Ishiguro registers the mortality which is screened from view by the driving business of the present, and which is part of what we will encounter if we stop and turn back. *Never Let Me Go* insists that we turn back nonetheless. Its ostensible theme of cloning, with its sci-fi connotations (and its intertextual reference to Aldous Huxley's *Brave New World*), might encourage us to read it as a dystopian novel about the future. However, it is explicitly set in the last decades of the twentieth century, in a borderland between past and present that prompts readers in the new millennium to consider the antinomies of 'progress'.

'Cloning' is a complex narrative metaphor. Metaphysically, it captures the paradox of our biological destiny, as unique beings whose evolutionary function is merely to pass on our DNA. Politically, it illuminates the contradictory nature of a society that claims to value each of its individual members, but then mocks such claims in the drab uses to which their lives are put. Kathy's peers never venture off the main roads; few of them, when they were being educated, imagined for themselves any future more rewarding than what is conjured up by one of their favourite pictures, a 'glossy double page advert... [showing] this beautifully modern open-plan office with three or four people who worked in it having some kind of joke with each other' (p. 131). Here, linked to the idea of what is 'modern', is an image – all the more disturbing because it represents the aspirations rather than the fears of these young people – of the individual's assent to a mode of life which (as Williams put the point in *Towards 2000*) extends 'beyond the basic system of an extraction of labour to a practical invasion of the whole human personality' (Williams, 1983, p. 261).

Ishiguro identifies a contradiction between the education of the individual and her subsequent sacrifice to the collectivity; but he does so in a spirit that is at least partly elegiac. *Never Let Me Go* offers, among its other meanings, an equivocal tribute to the equivocal, and threatened, achievements of twentieth-century social democracy. From the post-war period into the 1970s, Britain's social-democratic educational ethos upheld, even if only to thwart it, an ideal of free individual self-development. The closing sequence of the novel (p. 237ff) suggests that such egalitarian projects may have had their day. Kathy and Tommy are admonished: '"You were better off than many who came before you.

And who knows what those who come after you will have to face" ' (p. 243). The border territory in which the book is set provokes a half-regretful backward look, but opens no hopeful perspective on the future.

In *The Accidental*, a sense of the present as threshold or border is awakened by Amber/Alhambra, the 'accidental' visitor who turns up at the Norfolk house where the middle-class professional Smart family is staying. The Smarts, as their name suggests, are half-aware of the ills of contemporary consumer society, and at the same time deeply implicated in it, with their four-wheel drive car and 'their' cleaner. Amber is the troublesome spirit of questioning, opening up possible futures immanent in the present. The first sentence in which Eve and Michael Smart's 12-year-old daughter is named reads: 'Astrid Smart wants to know', and the first exchange between Amber and Astrid begins when the stranger throws the teenage girl an apple – the time having come for her to acquire knowledge, on the edge of adulthood (Smith, 2005, pp. 7, 31). This first chapter opens a threshold that corresponds both to Astrid's pre-pubescent, millennial sense of a beginning, and to the potential for new vision latent in her older brother, mother and stepfather. Much later, Amber will prompt Eve to acknowledge the questions she can no longer defer:

> You've led a life unthinkable to most of the generations of women and men who birthed you to freedoms and riches unimaginable to them ... And what is it they'd ask you, what is it you think they'd want to know, if they were here tonight, all those women and men and women and men and women and men that it took to culminate simply in the making of you, the birth of you ... ?
>
> (p. 182)

At a time when British literary fiction favours 'historical' pastiche (with 'mendacious glorified stories' of the Second World War particularly popular, as we are reminded: p. 82), Smith's novel is illuminated by its sense of historical choices to be made today and tomorrow.

The environs of the Smarts' rented holiday home, where Astrid wakes in the opening scene to record another summer dawn on 'the mini dvd tape in her Sony digital' (p. 8), are the setting for most of the action. This post-pastoral borderland foregrounds the destructive impact of the future on the past: the 'countryside ... is beautiful. It is really English and quintessential. [Astrid and Amber] watch the cars beneath them going in and coming out, moving like a two-way river. The sunlight off the windscreens and the paint of the cars is flashy in Astrid's eyes'

(p. 117f). They are returning from a walk through scrubby fields to the local supermarket, where they have eaten lunch in the car park, by 'a horrible recycling-bin place. They sit on the grass...in the smell of old wine and beer from the bottle bins. Our recycling project, a sign says by the bins. Success. Environment' (p. 116). 'Quintessential' rurality and environmental 'success': this highlights the fragility – or absurdity – of our invocation of aesthetic amenities and environmental criteria that we cannot really value, since we continue to destroy and ignore them.

The village itself is far enough from the roads to be quiet. In the 'substandard' house, grubby, as she sees it, with the detritus of old and dead people, Astrid fears that 'nothing is going to happen' (p. 8). Rural simplicity assumes its traditional pastoral power to make visible, and alien, the normally unremarked practices and encumbrances of the metropolitan everyday. ('Pastoral is "carnivalesque" in Bakhtin's sense of playfully subverting what is currently taken for granted: the hegemony of the urban...': Gifford, 1999, 22f.) The Smarts' holiday decouples them from the circuits of the contemporary, with its constant virtual presences and simulacra, and allows a re-engagement with sensory experience and unmediated time. Astrid's older brother Magnus, whose understanding of sex has been acquired from Internet porn sites, learns in Norfolk the enchantments of the flesh. As in Aldous Huxley's *Brave New World* (1932), the pornographic body signifies the seduction of the anonymous, the mechanical and the inauthentic. Smith's antidote is not the heroic (and misogynistic) asceticism of Huxley's John Savage, but the smells, dampness, intimacy and heat of fucking (p. 140ff). Magnus' lover wears a watch, but it always shows the same time (p. 143). Astrid, for her part, has left her mobile phone in London. Her digital camera, on which she has been filming everything, is flung from the motorway footbridge by Amber, without explanation or apology. Astrid is then forced to recognise that 'it is actually not true that not a single thing happen[s]' in the passage of an eventless, unrecorded minute (p. 126f). As in Shakespeare's Forest of Arden, where they 'fleet the time carelessly, as they did in the golden world' (*As You Like It*, I, i), a new appreciation of what lies around us is born out of radical pastoral simplification. Back in London, the Smarts discover that their house has been stripped of every furnishing, down to the stair carpets: it will be some while before they can resume civilised city life.

Return to the city is always implicit in pastoral, whose fictions are made by and for smart, metropolitan people. The pastoral scene is the scene of writing and art: an imagined space, a borderland, in which culture meditates on itself, on its distance from and dependence on non-cultural (or natural) origins and foundations. The novel's conclusion (p. 306), inviting us to consider the relation between desire, representation, consumption, and ecology, invokes this dialectic:

> Heaven on earth. Alhambra.
>
> It's a top-of-the-range but still-affordable five-door seven-seater people-carrier with a 2.8 litre engine that can go from 0–62 in 9.9 seconds.
>
> It's a palace in the sun.
>
> It's a derelict old cinema packed with inflammable filmstock. Got a light? See? Careful. I'm everything you ever dreamed.

Of the novels I have discussed, *The Accidental* has the most marked utopian dimension, or inflection. However, it does not propose any permanent 'return to nature'. It is to the city that we must return, having taken thought. If there is to be a golden age, it is yet to come.

Concluding reflections

In contemporary 'Western' first-world societies, increased consumption has become a primary goal and criterion of progress. Consumption is presented as the most rewarding of individual pleasures – with the possible exception of 'sex', if indeed sex is not just another form of or adjunct to commodified enjoyment. (*Platform* highlights this link between eroticised consumer pleasure and commodified eroticism, suggesting how both are bound up with some of the most cherished and important freedoms of the postmodern Western subject.) Most 'culture' in Britain today endorses this, in the commercial advertising which is among its best-funded and most pervasive forms, in much TV programming and journalism, and even in some literary genres (chick-lit; or the shopping-and-fucking novel, which Houellebecq mimics and subverts).

However, as Kate Soper argues in her Introduction, the contemporary moment is one of contradiction and denial. Governments seek to sustain the conditions for continuous growth, based to a large extent on expanded personal consumption. But they are also more and more speaking of the destructive consequences of that consumerist, growth-driven economy: acknowledging, belatedly, what environmentalists like Raymond Williams had understood 35 years ago – 'the ... decisive fact that we cannot *materially* go on in the old ways'. As I write these words (on 14 March 2008), European heads of government are meeting, once again, to issue warnings about the imminent dangers of climate change. On the same day, the Queen has been presiding over the opening of Terminal 5 at London's Heathrow Airport.

I have suggested how (in their quite different ways) Houellebecq, Ishiguro and Smith address this contradictory moment: by exploring some ethically problematic dimensions of contemporary consumption, by charting some of the bleaker landscapes it creates for us to live in, and also – this is the major emphasis, developing my introductory reflections on progress and retrospect – by referring to, invoking and reworking aspects of pastoral. My central contention, in terms of this book's focus on alternative pleasure, is that cultural traditions of pastoral, nostalgia and retrospect, when brought (as by these writers they are) into dialectical and imaginative connection with the new discontents of the contemporary, can still 'speak of something that is gone', and thereby imply a critique of what is.

Can such speech ever be more than elegy and valediction? Can its cultural imaginary inform a new political imaginary? Literary fiction, its readership confined to a small minority of the middle classes, appears most unlikely to bring about political changes, even modest ones: at best, it can be read as an expression of emergent critical sensibility – and 'cultural' expressions of distaste for the dominant mode of production have been circulating for as long almost as that mode has been dominant. My provisional conclusion is that novelists and poets may lament and warn as they like, but structural economic imperatives (the imperatives of globalised capitalism), in combination with the general public's continued appetite for more consumer goods and pleasures, will continue to constrain and largely to determine the public intellectual terms in which feasible futures are discussed.

Indeed, the novels, although they all have an implicitly didactic, consciousness-raising dimension, do not inscribe a very hopeful sense of their own likely effects. *Never Let Me Go* is openly disenchanted with the future that it envisages ('Who knows what those who come after you will have to face?'). Michel Renault is finely sardonic about the first-world consumer's pleasures of sex and travel, but he promotes them and enjoys them for all that. Amber's disquieting influence, once Michael and Eve have banished her from the Smart family, continues to affect the younger generation; but Astrid's schoolgirl gestures of allegiance to Amber's subversive vision hardly seem likely to lead to any real change in the world.

A less sceptical assessment can nonetheless be ventured. In a society centred on privatised consumption, it seems ever more difficult to organise or even to envisage the kinds of collective action and solidarity that were the basis of earlier attempts at transformative politics. However, it is also the case that the individual consumer, just because she or he has to go on consuming if the show is to stay on the road, has in principle a measure of political and indeed directly economic agency. This can become highly effective, once consumption comes to be understood as active rather than passive, and as the site of new kinds of intervention. That kind of effective, informed, self-aware consumption is the theme of several chapters here. It does not have to wait upon the creation of a large organised constituency. Cultural expressions of anxiety and disenchantment about the world made by consumerism can connect with, inform and develop individuals' emerging sense that their acts of consumption represent a moment of political agency.

References

Beer, Gillian (1988) *Can the Native Return? The Hilda Hume Memorial Lecture 1988* London: University of London.

Bronte, Charlotte (2004; first published 1853) *Villette* London: Penguin.

Byatt, A. S. (1985) *Still Life* London: Chatto.

Dickens, Charles (1996; first published 1860–1861) *Great Expectations* London: Penguin.

Emecheta, Buchi (1989) *Second Class Citizen* London: Hodder.

Garrard, Greg (2004) *Eco-Criticism* Abingdon and New York: Routledge.

Gifford, Terry (1999) *Pastoral* London and New York: Routledge.

Gissing, George (1985; first published 1892) *Born in Exile* London: Hogarth.

Hardy, Thomas (1995; first published 1878) *The Return of the Native* London: Everyman.

Hewison, Robert (1987) *The Heritage Industry: Britain in a Climate of Decline* London: Methuen.

Houellebecq, Michel (trans. Frank Wynne) (2003; first published in French as *Plateforme*, Paris, 2001) *Platform* London: Vintage.

Huxley, Aldous (1975; first published 1932) *Brave New World* Harmondsworth: Penguin.

Ishiguro, Kazuo (2005) *Never Let Me Go* London: Faber.

Kerridge, Richard (2001) 'Ecological hardy', in Karla Armbruster and Kathleen R. Wallace (eds) *Beyond Nature Writing: Expanding the Boundaries of Ecocriticism* Charlottesville and London: University Press of Virginia.

Lawrence, D. H. (1975; first published 1922) *Aaron's Rod* Harmondsworth: Penguin.

Lerner, Laurence (1972) *The Uses of Nostalgia: Studies in Pastoral Poetry* London: Chatto and Windus.

McGahern, John (2002) *That They May Face the Rising Sun* London: Faber.

McGahern, John (2005) *Memoir* London: Faber.
McIlroy, John (1993) 'Border country: Raymond Williams in adult education', in John McIlroy and Sallie Westwood (eds) *Border Country: Raymond Williams in Adult Education* Leicester: National Institute for Adult Continuing Education.
Orwell, George (1971; first published 1939) *Coming up for Air* London: Secker and Warburg.
Patterson, Christina (2006) Review of John McGahern's *Creatures of the Earth*, in *The Independent* (London) *Books and Arts Review* 15 December p. 22.
Ryle, Martin (1988) *Ecology and Socialism* London: Radius.
Ryle, Martin (2004) 'Surplus consciousness: Houellebecq's novels of ideas', *Radical Philosophy*, 126, pp. 23–32.
Shakespeare, William (1963; first published 1623) *As You Like It* New York and Toronto: Signet.
Smith, Ali (2006; first published 2005) *The Accidental* London: Penguin.
Wiener, Martin (1981) *English Culture and the Decline of the Industrial Spirit, 1850–1980* Cambridge: Cambridge University Press.
Williams, Raymond (1977) *Marxism and Literature* Oxford: Oxford University Press.
Williams, Raymond (1981) 'Response to the debate', pp. 142–152 in Martin Jacques and Francis Mulhern (eds) *The Forward March of Labour Halted?* London: Verso.
Williams, Raymond (1983) *Towards 2000* London: Chatto and Windus.
Williams, Raymond (1993; first published 1973) *The Country and the City* London: Hogarth.
Wordsworth, William (1965; composed 1802–1804) 'Ode: Intimations of Immortality…', in T. Crehan (ed.), *The Poetry of Wordsworth* London: University of London.

3
Ecochic: Green Echoes and Rural Retreats in Contemporary Lifestyle Magazines

Lyn Thomas

Simon Blanchard argues convincingly, in Chapter 4 of the present collection, that even apparently 'counter-consumerist' media representations such as 'Live 8' are enmeshed in neo-liberal ideologies and economic structures. Whilst agreeing with him that any sustained political challenge to the socially and environmentally destructive primacy of consumer 'needs' is unlikely to emerge from contemporary mainstream media, in this chapter I will explore other, more ambiguous aspects of the mediatised culture he describes. I will argue that even in the current context of commercial imperatives and neo-liberal hegemony, traces of counter-consumerist desires and of anxieties about the negative impacts of consumerist lifestyles can be found in mainstream media texts, where they might least be expected.

In the previous work on contemporary British lifestyle television, I have identified four sub-genres whose narratives are inflected with such anxieties and alternative desires (Thomas, 2008). I have described how in programmes focusing on relocation to rural or foreign environments the narrative is often triggered by an account of the disadvantages and constraints of high-speed urban living. Similarly, many cookery programmes exhort us to abandon commercially produced fast food in favour of wild, local or organic ingredients and home cooking. A number of celebrity cooks and gardeners have involved themselves in the politics of food production through their programmes: Jamie Oliver attempted to reform the school meals service while Hugh Fearnley-Whittingstall took groups of 'fast-food junkies' to his Dorset farm and taught them to cook 'real' food, and *Gardeners' World's* Monty Don tried to help a group of ex drugs offenders by teaching them to grow and cook food. If the quest for food produced in harmony with nature seems at times to have a spiritual dimension, a form of secular spirituality, values other than material success and acquisition are also sought through a turn to more mainstream religion in 'reality' television. For example, in a new hybrid of reality format and documentary, 'ordinary men and women' live for around six weeks in a religious setting, whether Benedictine Monastery, Convent of the Poor Clares or Muslim Retreat Centre. The first programme in these series, *The Monastery*, was framed by the Abbot, Father Christopher Jamison, as precisely a response to the spiritual void at the heart of contemporary consumerist society. Finally, a fourth sub-genre, which I have called 'eco-reality', confronts directly the environmental imperative to organise our domestic lives differently. BBC2's *It's not easy being green* showed the pleasures of gardening and of constructing an energy efficient and independent home, while *No Waste Like Home* showed that minor adjustments to domestic routines could reduce the piles of waste generated by most British households, and their energy consumption.

In this chapter, however, I am focusing on another area of lifestyle media – magazines – in order to explore the extent to which these glimmerings of disquiet and alternative desires might be colouring a broader cultural field. Given the interconnected nature of contemporary media in terms of production context, audience engagement, and intertextuality, it seemed likely that the concerns outlined above are not unique to television. Despite this, there are significant differences between the two media. Is it then possible to argue that lifestyle magazines, like lifestyle television, are concerned not only with conspicuous consumption, but also with desires that consumerism cannot satisfy? To what extent can the direct engagement with the damaging effects of affluent consumerism that we find in the 'ecoreality' series be found in a medium where there is no 'public service' mission, and in most cases less public monitoring of its ethical choices?

Which magazines?

The commercial underpinnings of lifestyle magazines are even more obvious than those of television in that the magazines consist almost entirely of direct advertising or 'advertorial', where a feature article is written in order to promote a service, shop or goods. In my sample of magazines collected in September 2005, in *Homes and Gardens* I found 37% direct advertising and 59% advertorial, in *Livingetc* 36% and 59% respectively, and in *Country Living* 48% and 27%. *Tesco* magazine was 33% direct advertising and 55% advertorial and Tesco promotions.

Livingetc and *Homes and Gardens* are typical of the genre in being almost 100% advertising, since advertising constitutes a major aspect of magazine revenue. *Tesco* magazine's main function is to promote loyalty among its customers, so direct advertising plays a slightly less dominant role. Even *Country Living*, which has the highest percentage of non-advertising content (25%), can hardly be described as free of capitalist imperatives. In fact as Robert Fish has observed, its articles often focus on people moving from city to country in order to create small but lucrative businesses (Fish, 2005); the magazine also gives awards for 'Enterprising Rural Women'.

A further sense in which these magazines can be seen as part of the cultural fabric of late capitalism is the fact that they emerge from a general move towards market segmentation, which began in the 1960s, and was taken up with increasing impact by the women's magazine industry in the 1980s (Gough-Yates, 2003). Each of the magazines I have selected has a distinct target reader, constructed to attract advertisers to the magazine and to reinforce its brand identity. For *Homes and Gardens* the reader is at an average age of 50 years, 'sophisticated, able to enjoy the finer things in life' and a 'classic, stylish, image-conscious home-owner' (see website 1, listed after the Bibliography: further websites are referenced by Figures 2, 3, and so on). For *Country Living*, the ideal readers are 'upmarket homeowners 35+ who either aspire to the romantic dream of living in the country or are already living the life' (2). *Livingetc* targets readers who are 'sophisticated, 30 something urbanites with great taste and a love of the good things in life. They are stylish, sexy, and so are their homes' (3). A spokesperson described the readership of *Tesco* magazine as 'mothers, 25–44 years old. In general it's the mid-market type of readership, those that read the *Daily Mail*, that fall under NRS category BCD' (Unpublished interview, 2005).

The *Homes and Gardens* readership is indeed predominantly middle-class, middle-aged, and female: according to the National Readership Survey, 68% fall into the ABC1 category, 60% are aged over 45 years, and 70% are women. Again for *Country Living* the readership is predominantly middle-class (70% ABC1), middle-aged (63% aged over 45 years), and female (68%). The profile of the *Livingetc* readership is very much younger (75% aged under 44 years), still more predominantly female (80%) and still more middle-class (81% ABC1) (4). Data on the *Tesco* readership is not provided by NRS, but as a free supermarket loyalty magazine, it is likely to attract a much broader readership in terms of age and class, whilst the predominance of female readers will be maintained. Inevitably the choice of lifestyle magazines focused on the home or on health and food as object of study here means that the representations I am discussing are targeted at women readers, and I will return to the issue of gender in my analysis of the texts. The four magazines selected offer some variety in that *Tesco* clearly addresses a much wider class and age base, while *Livingetc* is aiming at younger, more urban readers than either *Country Living* or *Homes and Gardens*.

In all four cases, the commercial nature of the media products I am analysing is further illustrated by their success in the market. *Homes and Gardens* and *Country Living* are established classics, achieving thirteenth and fourteenth place in the NRS listing of monthly women's magazines in 2007, and with readerships of 796,000 and 763,000 respectively (5). In 2005, *Livingetc* was a promising newcomer at fortyfourth place with 215,000 readers, and by 2007 it had consolidated its position, rising to fortieth place. *Tesco* magazine meanwhile, like other loyalty publications or 'consumer magazines', has achieved resounding success since its launch in 2004, with a reported readership of 2.5 million, and second place in the ABC listing, which is directed at advertisers (6). *Tesco's* success illustrates how free consumer magazines produced by supermarkets and other giant corporations play an increasingly dominant role in the lifestyle magazine market. Other more 'niche' lifestyle magazines, including those analysed here, are also produced by very large organisations. The apparent diversity of the magazine market is thus belied by the fact that magazines are produced by a very small number of large publishing companies, themselves owned by global corporations: *Livingetc* and *Homes and Gardens* by IPC, which is part of the Time Warner corporation, and *Country Living* by the National Magazine Company, owned by the Hearst Corporation. American capital and global corporate power are thus at the heart of these productions.

The 'greening' of lifestyle magazines?

To what extent, then, do such market-led and market-oriented productions reflect concerns about the negative environmental impact of high consumption, high-speed lifestyles? In September 2005, I surveyed the coverage of problems resulting from affluent, consumerist lifestyles in the West in a selection of lifestyle magazines: problems such as stress, work pressures, health issues, poor food quality, noise, pollution and congestion, and social inequalities. I also looked for signs of desires for pleasures other than those provided by the acquisition of consumer goods, pleasures which consumerism may even compromise or occlude. In the first part of this discussion I will focus on first of these two areas – the extent to which these magazines are registering and responding to problems resulting from social and economic structures based on profit (and hence on work and 'productivity'), and on high levels of consumer spending.

Tesco magazine carried 19% of relevant items at that time, with most focused on health and food, and one advertorial on the pleasures of woodland walks. By 2007, the now online magazine had developed a whole section on 'Going Green' (alongside 'Get Healthy', 'Lifestyle and fashion', 'Charity and Initiatives' and 'Recipe Ideas'). All of these sections have some relevance to the issues I surveyed in September 2005, while 'Going Green' directly addresses environmental problems. In July 2007, the 'Going Green' section included items on battery-powered delivery vans, reductions in the packaging of sandwiches, a website designed to help children count carbon footprints, the conservation of water and energy, recycling, and the Tesco 'Nature's Choice' scheme for farmers, which aims to promote certain environmentally friendly practices. It is clear that the need to display 'corporate responsibility' is now a driving force in the production of this magazine, and that environmental issues have to be addressed as part of this effort. Indeed, reading the magazine could lead one to conclude that Tesco exists only to promote practices such as recycling, energy and water conservation, fair trade, healthy living and the consumption of good, locally sourced food. Clearly, through this publication, Tesco is marketing itself as a responsible and ethical organisation, perhaps in response to criticism by environmental organisations, such as Friends of the Earth, and in the media (Monbiot, 2007; websites 7 and 8). In the 'Going Green' section, whilst some of the items suggest ways in which readers can live in a more environmentally friendly way, by saving rainwater or buying energy-efficient light bulbs (at Tesco's of course), most of the articles present Tesco's own achievements. This presentation is of course highly selective: a slight reduction in packaging on sandwiches is offset many times by the vast quantities of plastic enveloping other food products; a small number of battery-powered vans is outweighed by the armies of lorries delivering goods to Tesco stores throughout the world, and of cars driven to out of town superstores. But whilst these representations may emanate from corporate 'greenwash', they nonetheless represent an important socio-cultural change. The massive shift which has taken place in two years in Tesco's self-promotion through its magazine does indicate a response to a changed public mood – one where concerns about the origins of products, the food miles clocked up in bringing them to market and the working conditions of their producers are beginning to matter to consumers, and hence have to be taken into account by corporations. Even if Tesco's directors are cynically 'chasing the green pound', the change in direction by such a powerful organisation is significant (Finch, 2007, p. 32).

In September 2005 both *Homes and Gardens* and *Livingetc* contained almost no reference to environmental issues, and in September 2007 this situation was unchanged. The only significant exception to this is an advertorial in *Livingetc* of September 2005 on 'Ecochic Hotels'. Here readers are reassured that 'going eco does not mean forgoing style' as the magazine has found 'seven fabulous retreats with conscience-clearing credentials' (p. 138). What these credentials consist of remains hazy, though each hotel has some claim to ecologically aware practice, whether a grass-covered roof, resident ecologist or organic food. Throughout, the emphasis is on reassuring *Livingetc's* urban sophisticate readers that the hotels are eco-friendly *and* stylish, as it is clear that these characteristics are not considered to be compatible in the normal run of things: 'this award-winning "home from home" does planet-saving style without the Jesus sandals and alfalfa sprouts' (p. 140). The article in fact offers *Livingetc* readers an identity which is precisely *not* based on any serious environmental concerns, since this would inevitably lead to dire lapses of style such as the dreaded sandals. What is offered here is a fashionable gesture towards environmental awareness, in a lifestyle otherwise devoted to consumerism and its display: 'Show off your quirky-but-caring Brit style with a recycled outfit from Junky styling, which turns jumble sale cast-offs into sartorial miracles' (p. 140). Here it would be justifiable to speak not of greenwash, but of the palest of green politics as designer accessory. This conclusion is confirmed by the locations of the hotels – Botswana, Seychelles, Canada, Morocco – most of which can only realistically be reached by aeroplane. This aspect of the holidays proposed is of course not discussed.

Country Living, on the other hand, with its more country-oriented target reader, shows considerably more interest in environmental and social issues, which contribute 18% of the content of the September 2005 issue. Advertisements for charities such as Make Poverty History, recyclenow.com, and Macmillan cancer care interrupt the flow of promotion of consumer goods. Feature articles provide instruction on growing organically, making the home eco-friendly, running a smallholding or cooking home-grown vegetables. A fundamental narrative trope of the magazine is that of the move to the country, which as we have noted often involves setting up a successful small business. In this issue, 'The Whey Forward' told the story of a couple setting up a cheese shop in a small Cumbrian town (p. 70ff). Although this is without doubt a story of enterprise, the piece contains strong counter-consumerist elements, or more precisely elements which counter the current dominance of out-of-town superstores selling globally sourced foods: the shop as hub of community life, the promotion of local products, the restoration of an old building, the strengthening of the small town's high street, the modest profit margins.

Another feature article, 'Driving villagers round the bend' (p. 50ff), tackles the problem of rural traffic. Its narrative structure resembles what Fish identifies as the magazine's strategy for dealing with difficult issues, in which the problem and its solution are presented simultaneously, so that we have the sense that in the world of *Country Living* nothing is insurmountable: '[T]hey only acknowledge counter visions insofar as affirmative programmes of action are already in place' (Fish, 2005, p. 164). Nonetheless, articles such as this one on campaigns to improve rural road safety are addressing problems resulting from affluent lifestyles – in this case the proliferation of cars and the desire to travel at speed in all locations. One of the solutions proposed in the article is a 'Safe Lanes Drivers' scheme, which aims to reinstate the human dimension in the encounter between drivers and pedestrians: 'Now you make eye contact with drivers and there's a chance to say thank you for slowing down' (Nelson, 2005, p. 51). A second solution is traffic monitoring by residents using 'laser speed guns'. Clearly, neither of these strategies addresses the fundamental problems of excessive ownership and use of cars and poor rural transport infrastructures. They imply, like most lifestyle literature and broadcasting, that individuals can solve structural and political problems through their own choices and actions. The enterprising individual of the pages of *Country Living* is often a middle-aged, middle-class, country-dwelling woman. In this case, Jayne Bramwell (whose name itself speaks of apple trees and old England) is the instigator of the 'Safe Lanes' scheme: 'Sitting in her kitchen, with a sweeping view of the Sussex Weald before her, Jayne explains what prompted her to take action: "I felt vulnerable every time I rode my horse, Dazzle" ' (Nelson, 2005, p. 50). An image emerges of the country kitchen complete with Aga and large pine table, a pot of fresh coffee, and the country garden and stables outside. Even when dealing with threats to rural environments, *Country Living* presents an idyllic world; its heroines are well-meaning middle- and upper middle-class women who are both enterprising and caring. Because these women embody the aspirations of the magazine's target reader, and because of the magazine's rural focus, it cannot ignore environmental and related issues. Its engagement with them cannot be dismissed, and is clearly more serious than that of magazines oriented to urban readers, such as *Livingetc*. But in the end *Country Living* cannot make the difficult connections, for instance those between the middle-class incomers to the countryside whom it champions and the rising numbers of cars on rural roads or the house prices driving local young people out. It can argue for the donning of pink reflective clothing, but not for a campaign for a responsible transport policy. Nonetheless, within these limits, *Country Living* addresses aspects of environmental politics, which most lifestyle magazines ignore.

Escaping from the escape in contemporary lifestyle magazines

Perhaps the magazines I am analysing here in relation to counter-consumerism are most interesting in this context as representations of utopian lifestyles. In her research on 'alternative hedonism', Kate Soper has argued that consumerism not only threatens quality of life in terms of pollution, congestion, and so on, but also actually pre-empts and occludes the experience of 'simpler' pleasures – such as those of silence, of quiet contemplation, of time for relationships and social life, of contact with nature and the ability to roam freely, safely and unhindered by traffic in urban and rural landscapes (Soper 2007, p. 221ff). In this final section I will argue that the pleasures invoked by these lifestyle magazines are frequently of this order, even if consumer goods are portrayed as the means of achieving them. Amidst the glossy illustrations of desirable artefacts, and the advertorials thinly disguised as feature articles, a sub-text of desires for other pleasures and other ways of living can sometimes be unearthed. The articles I have chosen to analyse here, like the 'relocation' genre of lifestyle TV, are all concerned with the theme of escaping from the pressures of everyday life. In this sense, in very different ways, they present modern affluent lifestyles as something to be escaped from, and express a form of restlessness or dissatisfaction with lives premised on being busy and acquiring wealth.

In July 2007, one of *Tesco* magazine's lead features is on '10 Easy Steps to a hassle-free holiday'. An image of a couple lying in hammocks and gazing out to sea crystallises the desires for rest and relaxation in a peaceful natural environment, which the article might appeal to. The rest of the piece underlines how modern holidays in reality often fail to deliver this experience. Four of the nine points (one is repeated) are concerned with possible problems and ways of avoiding them: an overnight bag in case of flight delay, advice on family health from a government website, holiday insurance (from Tesco finance . . .) and an organisation to contact in case of a 'holiday nightmare'. All of the five remaining points are concerned with stress management. The first suggests that the pressures of the modern workplace make it virtually impossible to take holiday at all: 'at work, plan your escape two weeks before you go. Decide which tasks can wait until you return. Tell colleagues you cannot be contacted, brief them in detail and trust them to sort things out without you' (9). Others suggest the stress involved in the process of going on holiday itself; the holiday is likely to be so stressful that we are advised to 'put things in perspective', number our worries 1–10 and ignore those in the 1–5 category. Each family member should take with them a 'stress-soothing' kit. While on the beach, 'close your eyes, listen to the sea, smell the fresh air and feel the sun'. Then create a peaceful image based on how you feel – not for its own sake, or even to help you through the winter months ahead, but in case of a delay or problem on the journey home. Finally, avoid the panic of returning to an empty fridge by ordering food online from Tesco in time for your return. Thus the demands of domestic life reassert themselves in the midst of the holiday idyll.

This article indicates how dystopian modern affluent lives have become: the holiday is the high point of consumer success, the moment in the year when we are allowed momentarily to abandon the work ethic and experience unmitigated pleasure, in the process spending a large proportion of what has been earned. The article reveals the queue at the airport, the crying children, the pressure of packing for a family on top of finishing off work, the minor illnesses and fatigue of travel. It begins at work, and ends with domestic responsibilities, demonstrating to its women readers that there is no holiday from the second shift. Even the imagery is predominantly dystopian: a packed suitcase, a woman delving into a partially packed suitcase, and an emergency overnight bag containing the clear plastic containers required by airport security outnumber and outweigh the two images of beautiful beaches. The tone is pragmatic, and the advice seems to be aimed at women whose lives are constrained by limited resources and family pressures. There is no lament for lost pleasures here, just acceptance that this is how life is, and Tesco can help us cope: 'every little helps'.

For the more privileged readers of *Country Living*, escape is of a different order. In September 2005 the cover showed a painted Romany caravan surrounded by trees, sheep and carefully toning cushions and rugs. The headline is: 'ESCAPE to a dream hideaway' and sub-heading: 'Inspiring Ideas from . . . . A Romany Caravan, Railway Carriage, Designer's Hut and Garden Den' (*Country Living*, September 2005, pp. 20–29). The four spaces contrast masculinity and femininity. The caravan and railway carriage are decorated with pink floral cushions, stencilled wall decorations, lacy table cloths, vases of flowers, and antique bird cages. The designer's hut and garden den are plainer and more utilitarian; their occupants are male. The garden den has a brightly coloured checked blanket, 1960s melamine, tins of baked beans. All four 'homes' are connected by being photographed in the midst of nature, or with views beckoning out to the landscape. They speak of desires for a simpler, smaller version of the home, which is closer to nature, more mobile, and 'authentically' decorated with carefully chosen heritage pieces.

July 2005's *Livingetc* includes a piece on 'Decorating with Sheer fabrics', which moves from the use of sheer panels inside the home, to outside space, where you can 'keep your cool in the midday sun by draping a length of cotton muslin over your lounger' (p. 112) or 'create a private sanctuary by hanging a soft curtain panel across the exit to the garden' (p. 113). If this is not enough, the sheer fabrics, which 'look lovely when light streams through them', 'work beautifully as outdoor tablecloths' (p. 114). The remnants of a Mediterranean lunch are laid out on the tablecloth, indicative of the bonhomie generated by the creation of a halfway space, between inside and outside, predicated on a need to move beyond the enclosure of the home, to open up its boundaries. The article is about staying at home, embellishing the spaces of home, and using them to entertain guests, yet it is premised on travel, and the restlessness that inspires it. The colours – deep pinks and terracotta – that dominate the images are suggestive of exotic locations. The furniture is North African in style, and a pair of Moroccan slippers is casually placed by the bed. A silver teapot and glasses of mint tea adorn a coffee table in another image. The 'sheer fabrics' evoke the curtains and veils of the harem of the Western imagination. The 30-something urbanite's restless desires expand beyond the garden or meadow settings of *Country Living* into global appropriations, postcolonial collages. To shoot the images, the 'Livingetc team stayed at Caravanserai, Marrakech, Morocco' (*Livingetc*, July 2005, p. 115).

In July 2005, *Homes and Gardens* also carried the word 'ESCAPE' in pink capital letters on its front cover; the escape in this case was to 'four inspirational homes in France, Germany, Belgium and Australia', indicating its world-wide distribution and readership. Lower down, a second title 'Gone to the Beach: relaxed seaside living' reinforces the idea of escape. The titles are set against a background image from one of the 'inspirational homes': an interior painted white and pale blue leads the eye via a bucket of pale pink peonies to an open window framing a view of distant greenery. The 'Escape' headline is placed immediately above the window. Inside the magazine, all of the four homes featured reflect the theme of escape, whether from 'the humidity of the city to relax in the cool haven of [a] Mediterranean-style orangerie' (p. 55) near Antwerp; from the busy town centre to the calm of a Brittany apartment painted in 'mer du Nord' colours (p. 61); to 'a perfect combination of my job in town, which I love, and country life' in Northern Germany (p. 107); or from 'the rigours of city living' in Melbourne, to 'an idyllic rural retreat in the heart of southern Australia' (p. 93). The two-homes lifestyle which is the premise of three of these features (and of the Channel 4 *Relocation, Relocation* series) is presented in a still more extreme form in the first feature article in the magazine: 'This is where we hide from the world' (p. 32ff). This piece presents a family who divide their time between a Dutch barge on the Thames and an uninhabited island off North Wales, which they keep as 'as much of a private retreat as possible' (p. 35).

Like the *Tesco* magazine article, though in very different ways, these three pieces point to the increasing desirability and elusiveness of escape from the pressures of living and working in affluent, consumerist Western society. Whilst the representations here are based on significant wealth, with all the privileges and freedoms that this affords, the articles nonetheless speak of a never-ending search for peace and relaxation. A rural retreat is no longer enough; what is desired is an escape from the escape, or a retreat within the retreat. Thus the owner of the small manor house retreats to his orangerie, the *Country Living* rural property owners to dens and caravans in the garden, the *Livingetc* apartment dweller to a private sanctuary behind a voile curtain. In its most extreme form, even living on a boat is subject to urban pressures: 'The phone rings constantly and people drop by', necessitating the purchase of a deserted island 'where there is no phone and we're cut off from the mainland, we can hide away from the world and be a family together' (*Homes and Gardens*, July 2005, p. 35).

All of this might lead us to ask why the idea of escape is so seductive, and yet so difficult to achieve. These images of the lives of highly successful designers, explorers, artists, and so on present an upmarket version of the *Tesco* holiday dilemma. The high earnings exact a high price in terms of time for relationships and relaxation, and ever more convoluted solutions are sought. One might also comment that in these pages, the solutions sought always involve the accumulation of more consumer goods, and that a vicious, if beautifully decorated, circle is completed here. The 'escape from the escape', whilst requiring a new set of curtains and cushions, is nonetheless an attempt to step more lightly on the earth. The complex colour schemes, object-laden interiors and perfect four-course picnics with matching cutlery of the magazine pages lead quite logically to a desire for a shack by the sea or a meal of baked beans eaten off a tin plate. The magazines thus provide the antidote to their own excesses.

On the other hand, the complex arrangements of artefacts and design perhaps point nostalgically to a time when middle- and upper middle-class women had time to devote to aesthetic occupations and to the embellishment of their surroundings. This is particularly the case in *Country Living*, where prettiness and cosiness are the dominant aesthetic mode. As the reality of women's lives in all but the most privileged sectors is dominated by the stress of the second shift, the rush from the pressures of work to the demands of home, *Country Living* allows its readers to imagine a life where a nostalgic feminine aesthetics can be fully indulged, where vases of flowers can be refreshed each day, a cushion embroidered, a piece of lace laundered and ironed, a cake baked (Craig, 2007). Whilst *Country Living* takes the nostalgic mode further than the other magazines, they all share the feminine register, the emphasis on the gendered work of supporting relationships. The ideal spaces created in the magazines' pages are those of relationality; their subject is the mise-en-scène for human, and particularly familial, interactions. The address is to the woman reader who in the end is responsible for making it all go smoothly; the fantasy offered is that the picnic in the orchard with the perfectly co-ordinated rugs and cushions, or the Mediterranean lunch behind the white sheer fabric, will make it all work out, make everyone loving, kind and full of laughter. Even in the leisure world of lifestyle, women are required to multi-task, to create the perfect settings for the relationships they must maintain. Thus, in a further irony, the invocation of peace and relaxation is in reality an expression of the ceaseless labour of femininity, the endless striving for perfection or at least for the maintenance of the relational status quo.

Whilst reproducing a cycle of accelerated consumption, lifestyle magazines invoke many of the pleasures and values that the high-earning, high-spending lifestyle threatens: contact with unspoilt nature, peace, quiet, relaxation, and time with friends and family. It is not surprising that femininity remains the guardian of these small-scale, often domestic pleasures, and that the magazines' gendered address thus leads directly to the representations of forms of happiness which, in reality, consumer goods cannot provide. Analyses of advertising have long since identified the mobilisation of fantasies, which have no real connection with the items for sale (Lury, 1996; Williamson, 1978). What is of interest here is the expression, in the midst of utopian images, of a sense of dissatisfaction with the stresses and strains of lives devoted to earning and consuming, the need to escape, and to escape again.

Concluding reflections

My analysis of lifestyle magazines provides evidence of cases where political pressure to recognise the negative environmental impact of Western consumerism is beginning to take effect, at least in the mode of self-representation adopted by corporations. My argument also highlights the contradictions between these 'greenwashed' statements and images and the reality of building large out of town stores, or of flying blueberries from Chile or tomatoes from Israel to Europe or America throughout the year. Despite these contradictions, I would argue that efforts to present businesses as environmentally responsible, and the small changes in practice that ensue, though not a solution, are a step in the right direction and evidence of a general cultural shift. The example of *Livingetc*'s 'ecochic' hotels, on the other hand, demonstrates how in many areas of contemporary consumer culture 'green' practices are mobilised purely as lifestyle accessories. Even here though, and still more so in the images of escape discussed in my final section, a change in cultural consciousness can be detected. Whilst the magazines' producers may be cynically marketing to a new niche, the fact that they are doing so, and that this move is commercially viable, indicates that substantial numbers of readers now like their consumerism to be green-tinged.

As Kate Soper argues in the Introduction, there may be more political potential in the dissatisfactions and desires that underpin the need for multiple escapes from modern life that these magazines depict, than in ethical and altruistic motivations alone. The utopian images here invoke a 'simpler' life which is both closer to nature and more relaxed. The fact that they are enmeshed in consumerism through the implication that such pleasures can be obtained by having the right accessories for a picnic does not change the power of the images themselves, and the sense in which they portray lost and still deeply desired satisfactions. The question this raises for the future is whether a new imaginary can be developed, where we reconnect with the values and desires vehicled by media representations at the heart of consumerism, whilst recognising the potential to satisfy them without consuming much more than one's own time. If value could be attached to the pleasures themselves, rather than to their highly mediated forms, and if spaces could be created for their realisation, a significant and further cultural shift would occur. This would impact as much on the culture of work as on that of mediated pleasure and identity, since lack of time resulting from work-dominated lives is one of the main pressures to obtain pleasure through consumption. Contemporary consumers might more than ever before be living in one world while dreaming of another, and it is this disjunction, and its potential mobilisation, which might give us both room for manoeuvre and grounds for hope.

Acknowledgements

The research presented here was carried out as part of a project directed by Kate Soper on 'Alternative hedonism and the theory and politics of consumption', which was funded by the ESRC/AHRC Cultures of Consumption programme.

Websites consulted

1. www.ipcmedia.com/mediainfo/homesandgardens
2. http://www.natmags.co.uk/magazines/magazine.asp?id=14
3. www.ipcmedia.com/mediainfo/livingetc
4. http://www.nrs.co.uk/open_access/open_topline/women/index.cfm(consulted 10 October, 2005)
5. http://www.nrs.co.uk/open_access/open_topline/women/index.cfm(consulted 12 August, 2007)
6. http://www.apa.co.uk/cgi-bin/go.pl/news/article.html?uid=1187(consulted 12 August, 2007)
7. http://www.foe.co.uk/ resource/ briefings/ the_tesco_takeover.pdf (consulted 12 August, 2007)
8. http://www.guardian.co.uk/supermarkets/story/0,,1996504,00.html(consulted 20 August, 2007)
9. http://www.tesco.com/todayattesco/0707_04_146_tg.shtml

References

Craig, Lyn (2007) 'Is there really a second shift and if so who does it? A time-diary investigation' *Feminist Review*, 86, pp. 149–170.

Finch, Julia (2007) 'Tesco wins battle for garden centre chain' *The Guardian*, 18 August.

Fish, Robert (2005) 'Countryside formats and ordinary lifestyles', in David Bell and Joanne Hollows (eds) *Ordinary Lifestyles: Popular Media, Consumption and Taste* Maidenhead and New York: Open University Press.

Gough-Yates, Anna (2003) *Understanding Women's Magazines: Publishing, Markets and Readerships* London and New York: Routledge.
Lury, Celia (1996) *Consumer Culture* Oxford and Cambridge: Polity Press.
Monbiot, George (2007) 'If Tesco and Wal-Mart are friends of the earth, are there any enemies left?' *The Guardian*, 23 January (see 8 above).
Nelson, Francesca (2005) 'Driving villagers round the bend', *Country Living*, September.
Soper, Kate (2007) 'Rethinking the "Good Life": The citizenship dimension of consumer disaffection with consumerism', *Journal of Consumer Culture*, 11(29), pp. 205–229.
Thomas, Lyn (forthcoming in 2008) 'Alternative realities: Downshifting narratives in contemporary lifestyle television', *Cultural Studies*, 22 (5–6).
Williamson, Judith (1978) *Decoding Advertisements: Ideology and Meaning in Advertising* London and New York: Marion Boyars.

Filmography

Growing out of Trouble (2006) BBC2.
Jamie's School Dinners (2005) Fresh One productions for Channel 4 .
Gardener's World (1969–) BBC2.
It's not Easy Being Green (2006, 2007) BBC Bristol for BBC2.
No Waste Like Home (2005) Celador for BBC2.
Relocation, Relocation (2003–) IWC Media for Channel 4.
The Monastery (2005) Tiger Aspect for BBC2.
The River Cottage Treatment (2006) Keo Films for Channel 4.

4
Mediated Culture and Exemptionalism

Simon Blanchard

This chapter is in six sections. I start with the last word in my title, outlining its emergence and relevance. I then review what has been called the mediatisation of contemporary consumer culture, and television's central role in that process. The third section suggests that we recognise television is a realm of ideas, and that what I call the telegentsia has been an important intellectual stratum of capitalist democracy since the 1950s. The fourth section looks briefly at television's history as an intellectual field, and as a platform for the diffusion of neoliberal ideas and the 'market society' ethos. The fifth section reviews a relevant case study – the G8 'Live 8' project. I argue that 'Live 8' illustrates the role of the telegentsia in sustaining an exemptionalist outlook. I further suggest that this event is an example of a wider mediated culture phenomenon that I call 'popwash'. After reviewing my argument thus far, I finish with some concluding reflections on media politics, the 'new wave' of challenges to exemptionalism, and the more long run, epochal dilemmas facing mediated 'consumer' society.

Exemptionalism

In 1997, in a collection of essays called *In Search of Nature*, the Harvard conservationist Edward Wilson set out a widely cited contrast between two rival ways of thinking about the prospects for the 'consumer' society. Wilson suggested that the contemporary epoch was notable for the clash between what he termed *exemptionalism* and *environmentalism* (Wilson, 1997). As a world view, exemptionalism holds that humankind is *not* subject to the ecological constraints that bind all other species, and can transcend natural limits by the use of human ingenuity, technical innovation, 'market forces' and so forth. Conversely, environmentalism sees humankind as a biological species dependent on – and emphatically part of – the natural world, and one whose long-run prospects depend on responsible stewardship of its planetary habitat.

In retrospect, what was striking about Wilson's schema was not its identification of environmentalism as an intellectual paradigm, conventionally dated in its modern form from Rachel Carson's *Silent Spring*, published in 1962 (Carson, 1962; McCormick, 1995, p. 55ff). Rather, what Wilson's argument brought more sharply into view – by the act of naming it – was environmentalism's less visible (but arguably more influential) rival outlook. Wilson's contrast – first aired in a *New York Times Magazine* article in 1993 (Wilson, 1993) – had been set out against the backdrop of the United Nations Conference on Environment and Development in Rio de Janeiro in June 1992, an event which had provided at least some grounds for thinking that the environmentalist agenda was enjoying renewed momentum. Against this mood of optimism, Wilson was at pains to stress the continued dominance of the deep background assumptions of the exemptionalist agenda, which was underpinning accelerated species loss, habitat destruction and growing climate turbulence – trends all highlighted in his landmark study *The Diversity of Life* (Wilson, 1992).

Mediated culture and 'the TV'

A decade later, the struggle between these two rival world views has arguably reached a new level of intensity, and Wilson's schema remains of great value in clarifying the challenges facing the 'consumer' society. In particular, I would argue that it helps us to identify the extent to which contemporary Anglophone consumer culture is profoundly, even tragically, exemptionalist and to recognise that what has been called the 'mediatisation' of consumption is at the heart of that process. What, then, is 'mediatisation' ?

We can begin to map 'mediatisation' as a phenomenon by noting first that 'the media' and 'mediation' as a social-cultural force have been with us for a long time. Although there were earlier explorations of this theme (for example Fisher, 1948; Veblen, 1899), a new current of historical scholarship emerged in the 1970s which has examined in great detail the 'media' dimension of the rise of 'consumer' society across the seventeenth and eighteenth centuries, the growth of advertising, and so forth (Berg, 2005; Brewer and Porter, 1993; Plumb, 1973). 'Mediatisation', however, refers to the more recent epoch (roughly since the 1950s) in which the media nexus takes on a far more pervasive and intensive character, such that scholars have argued that these two arenas – media and the growth of 'consumer' society – have become effectively fused.

As recent studies by Jannson, Louw, De Zengotita and others have argued (De Zengotita, 2005; Jansson, 2002; Louw, 2005), we now live in a media-saturated culture in which, as Andre Jannson has put it, 'media culture and consumer culture are theoretically overlapping and empirically inseparable categories', resulting in a life-world in which 'the everyday media context actually constitutes an integral part of socio-cultural processes, rather than something external to them' (Jannson, 2002, p. 11).

Taking the 'mediatisation' thesis one step further, I would suggest that the pivotal 'engine' of this epochal process has been the diffusion and expansion of television over the last half century. As is well known, 'the TV' has come to occupy a pre-eminent place at the heart of consumer culture. Taking just the British case, the number of households in the United Kingdom with at least one 'TV' rose from 5.7 million in 1956 to 25.2 million by 2006 (Broadcasters' Audience Research Board, 2007, *Television Ownership*). By this time the consumption of television provided a cornerstone of daily life rivalled only by sleep and work: average weekly viewing per person in 2006 fluctuated seasonally within a range from about 22 hours in July to around 27.5 hours in December (Broadcasters' Audience Research Board, 2007a, *Viewing Summary*). The weekday audience reaches a peak around nine o'clock in the evening, when the Office of Communications (Ofcom) reports that in 2005 there were an average of 23.4 million people watching television (Ofcom, 2006, p. 75).

Looking back, we can see that it was awareness of the early stages of these epochal shifts which marked out the intuitions found in those three classic commentaries from the 1960s, Boorstin's *The Image* (1962), McLuhan's *Understanding Media* (1964) and Debord's *The Society of the Spectacle* (1967) – a dissenting trilogy which sketched out the contours of the new 'image society' that was taking shape on both sides of the Atlantic (Boorstin, 1962; Debord, 1967; McLuhan, 1964; Tunstall and Machin, 1999).

The telegentsia

If we acknowledge how central television has been a driver of mediated consumption, then I would suggest this implies two related propositions. The first is that television is a realm of ideas, or what the French sociologist Pierre Bourdieu called an 'intellectual field' (Bourdieu, 1971, 1998). Second, we need to register that the custodians of that terrain are an intellectual stratum, which I shall call the 'telegentsia'.

If we acknowledge the telegentsia's existence, it provides us with at least one possible answer to the question Frank Furedi posed in his recent polemic *Where Have All The Intellectuals Gone?* Of course, the argument that television provides a central place for ideas and intellectuals would not satisfy Furedi, who is clearly committed to a strongly normative model of 'intellectuals'. For Furedi, intellectuals are by definition dissenting: critically and socially at odds with the society around them. However, as Stefan Collini and Thomas Heyck have reminded us (Collini, 2002; Heyck, 1998), we need to be very wary of assuming that the 'symbolic aristocracy' is typically or necessarily contrarian. Equally, much of the prevailing literature on intellectuals has been built on a pre-twentieth century vision of them as – literally – putting pen to paper, and this has meant that commentary on them has been rather slow to acknowledge that they can be found in large numbers at the heart of the 'image society'. I would suggest that we must indeed recognise what we can call television's 'ideational' role (Blyth, 1997, 2002; Hay, 2002, p. 251ff) and its central part in sustaining the doctrinal and 'cosmological' assumptions of mediated consumer culture.

If television has an ideational role, this raises questions about what kind of ideas can be found there. What has been the prevailing 'world view' of the telegentsia? How might we characterise this intellectual field? In answering these questions, we need to take account of television's social 'embeddedness' (Granovetter, 1985): the varied ways in which the telegentsia are both *shapers of* and *subject to* the wider intellectual climate of the times.

If we look at the overarching trajectory of British television, we find that its reigning ethos from the late 1940s to the 1970s was broadly liberal-centrist and doctrinally conformist. Its ideational contours were generally in tune with the wider textures of 'middle opinion' and Labour-Conservative 'mixed economy' thinking (Harrison, 1999; Oliver and Pemberton, 2004). Perhaps the most characteristic expression of this ethos can be seen in the élite traditionalism of Kenneth Clark's 13-part ruminations on *Civilisation*, broadcast by the BBC in 1969. That said, television at this time did provide some recurrent space for dissenting and left-reformist thinking in drama, documentary, arts commentary and such like. Examples of this would include the 'satire boom' of the early 1960s, the early work of Ken Loach and Tony Garnett, ITV's *World In Action* and John Berger's *Ways of Seeing* (Berger, 1972; Carpenter, 2002; Crisell, 1991; Fuller, 1998; Goddard, 2004). Arguably, this heterodox current of ideas within the telegentsia reached a post-war peak at Channel Four in the 1980s. During the tenure of Jeremy Isaacs, the Channel's first Chief Executive, significant regular space was found for openly 'New Left' and explicitly counter-cultural ideas (Blanchard and Morley, 1982; Darlow, 2004; Isaacs, 1989).

Nonetheless, as Blyth, Cockett and Harvey have shown, from the 1970s the wider intellectual climate amongst political and business élites was shifting sharply to the right (Blyth, 2002; Cockett, 1994; Harvey, 2005), and the telegentsia played a significant role in articulating and legitimising this newly emergent neoliberal 'common sense'. In this intellectual frame it was 'obvious' that the post-war 'welfare' settlement could not be sustained, that union activism was the root cause of present discontents, and that the prescriptions for crisis resolution offered by the increasingly vocal New Right offered the best – indeed the only – way forward. As the Glasgow University Media Group showed in their landmark 1976 study *Bad News*, this neoliberal frame was already prominent within television news and current affairs by the mid-1970s, and it entrenched itself further after the 'right turn' of the Thatcher governments.

Television since the 1970s: The telegentsia turning right

What has happened to the institutional and ideational contours of television since the emergence of the New Right? We can identify two broadly inter-connected processes at work: the 'marketization' of television's institutional structure and an equally emphatic 'market populist' re-making of its programme schedules and genres. We can resume these two processes in brief outline.

To begin with, the New Right has facilitated the ongoing expansion of television as an economic-industrial sector, a dynamic that can be seen most clearly in the dramatic growth of satellite, cable and digital channels in recent years, particularly since the late 1980s (Barnett *et al.*, 2000; Goodwin, 1998). Furthermore, this expansion has taken the form of a very deliberate shift towards more explicitly 'market' norms and mechanisms. As a result, television is now much more emphatically and assertively market-driven, with an aggressive emphasis on ratings maximisation, market populism and programme profitability (Leys, 2003). This new dynamic has been underpinned by another critical component: the phased (and ongoing) expansion of 'outsourced' programme production by broadcasters to the commercial 'independent' sector (Darlow, 2004; Woodward, 1990).

As a consequence, television as a vocational arena has now become polarised between a relatively small and increasingly wealthy élite of senior executives, 'name' presenters, television 'personalities' and suchlike, and a wider majority of the workforce, most of whom are now employed on a more casualised and short-term contract basis (Barker, 2005; Darlow, 2004a, 2006).

Secondly, the neo-liberal world view has gradually come to dominate not just thinking about television's institutional structures and regulatory frameworks, but also its programme ideas, thematic preoccupations and genres. As a range of commentators have noted, television schedules are now suffused by a pervasive market populism in which the touchstones of the neoliberal world view – the joys of 'enterprise', hyper-individualism, celebrity worship, money as the supreme arbiter and so forth – are played out and normalised (Dovey, 2000; Palmer, 2003). We can also see these processes at work in the way that the schedules have been re-drawn. Schedules provide the central organising framework through which both the telegentsia and the audience 'make sense' of television, and by which generic hierarchies, relative intellectual standing and credibility (or its opposite) are implied and orchestrated (Ellis, 2000). As the telegentsia has embraced neoliberalism, schedules have been recast to expand and foreground 'entertainment' genres and modalities, and relegate dissonant factual and political material to the outer reaches of the schedules and to smaller audience channels.

What has emerged from this is an intellectual field which is notably paradoxical. On the one hand, television is now an arena which goes to great lengths to *avoid* the signs of abstract, conceptual frameworks, seeking instead to wrap everything in an 'accessible' and 'entertaining' market populism. Even so, this arena is undoubtedly one with a very emphatic ideational agenda, in which audiences are to be relentlessly disciplined and 'coached' in the challenges of this new 'reality', and the requirements and 'common sense' of the new market order (Bell and Hollows, 2005; Moseley, 2000; Palmer, 2004).

The cumulative effect of all these changes has been to shift television towards a far more 'liberalised' and entertainment-led ecology, in which the inherited framework of 'public service' ideals is now largely residual, and the policy touchstones are 'choice', 'competition' and 'light touch' regulation – a new dispensation which was consolidated with the launch of the Office of Communications (Ofcom) in 2003. The long-run impact of change on the ideational commitments of the telegentsia has also been considerable. As the New Right's market 'restoration' gathered speed in the late 1980s, the inherited balance of forces between a majority liberal centrism and a minority strand of dissident 'leftism' began to unravel, a process symbolised by the departure of Jeremy Isaacs from Channel Four in the late 1980s (Isaacs, 2007). Over the next decade the political centre of gravity within the telegentsia began to shift more emphatically to the right. A new generational cadre of 'market friendly' commissioning editors, independent producers and policy 'consultants' came to dominate the institutional landscape of television (Darlow, 2004; Preston, 2003), gaining added traction and ideological fervour from the wider climate of market triumphalism which followed the collapse of the Soviet bloc. This process was then further accentuated by the ideological contortions of the so-called 'War on Terror', and by the Blair government's protracted battles with (and political victories over) the BBC concerning the reportage and framing of the invasion and occupation of Iraq (Dyke, 2004; Oborne, 2007).

By the time that Tony Blair left office in summer 2007, the intellectual remaking of the telegentsia since the late 1970s was largely complete. As a result of these changes it was now liberal centrists who formed the 'left' edge of televisual opinion, and these voices were now an increasingly residual minority tradition within an ideational spectrum which was led from and dominated by the Right (Alibhai-Brown, 2007; Edwards and Cromwell, 2006; Hari, 2007; Kampfner, 2005; Paxman, 2007).

'Live 8', or television + pop + neoliberal politics = popwash

The above highly compressed sketch of the telegentsia's right turn since the 1970s, and its growing embrace of neoliberal ideas, leaves us with at least one analytical conundrum: How do we make sense of television's role as an arena for regular outbreaks of on-screen 'caring and sharing', such as the Group of Eight (G8) 'Live 8' saga of Summer 2005?

We can begin to resolve this apparent contradiction by noting that Live 8 was part of a broader array of media-political ventures tied into the G8 process and the 2005 G8 summit, including the work and report of the British-government sponsored 'Commission for Africa' and the broader 'Make Poverty History' campaign (Payne, 2006; Williams, 2005, 2005a). A range of commentary has been sharply critical of the assumptions and agendas which underpinned this 'moment' and has drawn attention to the political conformism and deeply compromised gesture politics which shaped both its emergence and outcomes (Bond, 2006; Brutus and Setshedi, 2005; Hodkinson, 2005, 2005a, 2005b; Quarmby, 2005). Evaluations of the G8 summit and the Live 8 offshoot suggest strongly that they are likely to contribute nothing of consequence to reducing global inequality, or to meeting the drastic challenges facing the African sub-continent (Bond, 2006; Cammack, 2006; Payne, 2006; Porteous, 2008; Williams, 2005, 2005a).

However, if we acknowledge the force of these critiques, some alternative framework is still required to account for events of this kind. If they do not act as significant catalysts of positive change, what role do they fill?

We might start an alternative analysis by looking at Live 8's promoters: the people who made it happen. Its prime movers – Bono, Bob Geldof, Richard Curtis, and suchlike – are all very wealthy members of what Leslie Sklair has called the transnational capitalist class, firmly embedded in élite social movements and networks (Sklair, 1997, 2001; Snoddy, 2006). It has been through élite networks of this kind that much of the long-run diffusion of the neoliberal agenda has taken place (Levi-Faur, 2005).

We might also note that these events were prime examples of what Patrick West has termed 'conspicuous compassion' (West, 2004): ostentatious but ineffectual displays of moral rectitude which provide 'feel-good' emotions at very low personal cost. In other words, the romance of 'radicalism' without the risks of significant systemic challenge. For the promoters, Live 8 also provided an opportunity to reinforce their status and public profile, and to reinvigorate their 'sainthood' in the warm glow of the telegentsia's attentions.

Seen in this more sceptical light, Live 8 was a classic example of televisual exemptionalism, which focused the media frame for the G8 meeting on neoliberal critiques and 'solutions', and kept media and public attention away from those groups and ideas around the summit which were seeking to contest its purpose, vision and objectives (Bello, 2004; Hubbard and Miller, 2005; Harvie, Milburn, Trott and Watts, 2005). As with the annual 'Red Nose Day', Live 8 offered a politically conservative and deeply orthodox representation of global politics, one which posed no significant challenge to élite agendas or concerns, and afforded no central place to those directly affected – except as victims and supplicants. It was therefore hardly surprising that it produced no substantial benefits for those it was claiming to speak for. As a typical example of the harnessing together of the telegentsia, commercial popular music and élite political agendas, it provided a consumer spectacle which harnessed anti-systemic energies and diverted them along safe, corporate-friendly and pro-systemic pathways.

If we seek the media-generic inspiration for this type of event/process, then I would suggest that it can be found in the rise of corporate public relations (Davis, 2002; Dinan and Miller, 2007; Miller and Dinan, 2008), and, more particularly, the modes of this in what has been called 'The Age of Greenwashing' (Athanasiou, 1997, p. 227ff). According to Karliner's capsule history of this phenomenon (Karliner, 2001), 'greenwash' emerged in the mid- to late 1960s as a direct corporate response to the environmentalist agenda, offering an emphatically commercial and wholly exemptionalist response, but dressed up in 'green' slogans and imagery. As Rowell and Switzer have shown in detail, by the mid-1990s this type of 'corporate environmentalism' was a complex, well-funded 'backlash' movement on a global scale (Rowell, 1996; Switzer, 1997), and it continues to deploy its varied repertoire of exemptionalist 'solutions' through to the present day (Beder, 2002, 2006; Goodin, 1994; Monbiot, 2006). By analogy with 'greenwash', we might reasonably name the blend of pop, television and élite neoliberalism 'popwash' (Blanchard, 2007b).

Concluding reflections

To sum up my argument thus far: I have suggested that contemporary 'consumer' society is profoundly mediatised, and that TV and the telegentsia have been for roughly half a century the central 'engine' of this historically novel 'image society'. Moreover, I have suggested that the doctrinal ethos and world-view offered by the telegentsia has become increasingly attuned to (and assertive of) prevailing neoliberal orthodoxies, and that – in Edward Wilson's term – these orthodoxies are deeply and tragically exemptionalist. I have further suggested that the telegentsia play a critical role in events such as Live 8, which harness genuine public concern about the very real global challenges we now face, and re-direct this into inconsequential and pro-corporate forms of consumer spectacle. These forms present the exemptionalist agenda in a fundamentally fraudulent 'caring' and 'environmental' packaging, in which the central role of commercial popular music suggests the coining of a new critical term: 'popwash'.

In the light of this argument, what broader reflections are suggested by its themes? I shall touch briefly on three contexts: 'post-democracy' and the demise of the 'fourth estate'; the role of new media networks in contesting exemptionalism; and the twilight of fossil fuel modernity.

First, the inherited assumption in liberal-democratic argument about the separation of powers between 'politics' and 'media' now looks increasingly implausible, and has been subject to a growing body of critique in recent years. Several commentators have explored the 'mediatisation' and 'restyling' of party politics, public policy, statecraft and civic action (Blumler and Kavanagh, 1999; Corner and Pels, 2003; Meyer, 2002; Oborne, 2007). In its more assertive formulations, this argument contends that we are now moving (or may have already moved) across an historical threshold into a novel type of political regime, one which Colin Crouch has termed 'post democracy' (Crouch, 2004). In this regime, the inherited pluralisms and countervailing powers of democratic politics are being dismantled and subverted. In their place we now find a statecraft of 'manipulative populism' (Oborne, 2007, pp. 297–309) in which the 'media class' and the 'political class' are increasingly convergent, sharing joint responsibility for the overarching neoliberal project. Moreover, the evidence to hand suggests that this hybrid media-polity has not brought renewed lustre to the political arena: quite the reverse (Hay, 2007). Instead, this convergence process has *reinforced* patterns of declining legitimacy and public trust across *both* arenas, contributing to a 'legitimation crisis' that now encompasses both halves of this novel 'mediated governance' system (Mair, 2006).

Such an analysis suggests that we would be unwise to expect much in the way of serious challenges to the exemptionalist agenda from within the televisual heartlands of mediated culture. The telegentsia's leading cadres have been major beneficiaries of neoliberalism, and we cannot credibly expect them to make sustained space for the exploration of ideas and outlooks which challenge that worldview. To be sure, there will be moments of dissent, as well as recurrent outbreaks of 'environmentalist' greenwash and popwash – like the Summer 2007 'Live Earth' events (Marshall, 2007); but these momentary discords will not disrupt the telegentsia's enthusiastic embrace of consumerism-as-usual. Nor will they redress the atmosphere of banality and reaction that now pervades large swathes of the television schedules.

Thankfully, these trends have not gone unchallenged. Over the last decade or so, we have begun to see the emergence of a new wave of civic media activism and ideas *outside* TV, offering a dissenting, networked 'counter power' to prevailing orthodoxies (Benkler, 2007; Blanchard, 2007a; Castells, 2007; Waltz, 2005). Two examples of this might be noted here. The first is what amounts to a documentary film and video 'renaissance' – embracing cinemas, DVD, websites and festivals – which offers a new social context for the circulation of progressive voices and insights. This cultural network/channel has promoted incisive critiques of corporate power (Achbar and Abbott, 2005; Armstrong and Loach, 2006; Gibney, 2007; Greenwald, 2006), and provided space for thoughtful meditations on energy use and environmental politics (Gelpke and

McCormack, 2008; Greene, 2004; Guggenheim, 2006; Morgan, 2006). The second example is the expanding nexus of progressive websites offering inspiration, ideas and contacts for those challenging the exemptionalist agenda (For example, Energy Bulletin, 2004; Global Public Media, 2001; GreenTv, 2006). Seen in a broader historical context, this 'new wave' can be placed as a central facet of the wider network/culture of new social movements that has emerged to contest the neoliberal agenda (Evans, 2005; Merchant, 2005; Mertes, 2003). As matters now stand, it seems clear that it will be from these movements and voices that we will find better ways to live: ways that do not sacrifice our habitat and human prospects to the delusions of exemptionalism.

Finally, one much broader 'long view' context suggests itself here, as we try to make sense of the dangers and complexities we face. If we are to find our way through what Kunstler has memorably termed the 'Long Emergency' (Kunstler, 2005), we need to recognise that contemporary 'turbo consumerism' represents something close to the final phase of a very long historical arc of industrialised consumption patterns: the twilight of an extended Consumer Epoch now entering its protracted 'end game'. As Tony Wrigley has shown (Wrigley, 1988; 1992), the improbable transition to industrialised modernity has been underwritten by the legacy of fossil fuels. The accelerated hyper-consumption which now characterises the rhythms and routines of corporate, neoliberalised 'image societies' is a more frenetic, more spectacular version of this life-world, a late modern coda in exaggerated form. In this perspective, the central challenge that lies ahead will be to devise a further – perhaps equally improbable – transition: this time *beyond* the parameters of 'fossil fuel modernity' (Hopkins, 2008; Kunstler, 2005, pp. 22–60; Wackernagel and Rees, 1996).

References

Achbar, Mark and Abbott, Jennifer (2005) *The Corporation* [DVD] London: Metrodome.

Alibhai-Brown, Yasmin (2007) 'BBC "open to right wing populism"' *Media Workers Against the War* 23 May (consulted on 2 June 2007), online at: http://www.mwaw.net/2007/05/23/bbcmovesright.

Armstrong, Franny and Loach, Ken (2006) *McLibel* [DVD] London: Revelation Films.

Athanasiou, Tom (1997) *Slow Reckoning: The Ecology of A Divided Planet* London: Secker and Warburg.

BARB (2007a) 'Monthly Viewing Summary, July 06–December 06' *Broadcasters' Audience Research Board,* online at: http://www.barb.co.uk/viewingsummary (consulted on 6 June 2007).

Barker, Alex (2005) 'Number of top BBC staff paid over £100,000 soars' *Financial Times*, 23 December.

Barnett, Steven, Bottomley, Virginia, Cave, Martin and Graham, Andrew (2000) *E-Britannia: The Communications Revolution* Luton: University of Luton Press.

Beder, Sharon (2002) *Global Spin: The Corporate Assault on Environmentalism* Totnes: Green Books.

Beder, Sharon (2006) *Suiting Themselves: How Corporations Drive the Global Agenda* London: Earthscan Publications.

Bell, David and Hollows, Joanne (eds) (2005) *Ordinary Lifestyles: Popular Media, Consumption and Taste* Maidenhead: Open University Press.

Bello, Walden (2004) *Deglobalization – Ideas for a New Global Economy* London: Pluto Press.

Benkler, Yochai (2007) *The Wealth of Networks: How Social Production Transforms Markets and Freedom* London: Yale University Press.

Berg, Maxine (2005) *Luxury and Pleasure in Eighteenth Century Britain* Oxford: Oxford University Press.

Berger, John (1990; first published 1972) *Ways of Seeing* Harmondsworth: Penguin.

Blanchard, Simon (2007a) 'Media activist initiatives', in Kate Coyer, Tony Dowmunt and Alan Fountain (eds) *The Alternative Media Handbook* Abingdon: Routledge.

Blanchard, Simon (2007b) *Popwash: Reflections on Rock and Rule* Unpublished paper.

Blanchard, Simon and Morley, David (eds) (1982) *What's This Channel Fo[u]r* London: Comedia.

Blumler, Jay and Kavanagh, Dennis (1999) 'The Third Age of Political Communication: Influences and Features' *Political Communication,* 16, pp. 209–230.

Blyth, Mark (1997) 'Any More Bright Ideas? The Ideational Turn of Comparative Political Economy' *Comparative Politics* 29 (2), pp. 229–250.

Blyth, Mark (2002) *Great Transformations: Economic Ideas and Institutional Change In The Twentieth Century* Cambridge: Cambridge University Press.

Bond, Patrick (2006) *Looting Africa: The Economics of Exploitation* London: Zed Books.

Bond, Patrick, Brutus, Dennis and Setshedi, Virginia (2005) 'Average white band' *Red Pepper* (July 2005; consulted on 2 June 2007), online at: http://www.redpepper.org.uk/global/x-jul05-whiteband.htm.

Boorstin, Daniel (1992, first published 1962) *The Image: A Guide to Pseudo-Events in America* New York: Vintage Books.

Bourdieu, Pierre (1971) 'Intellectual field and creative project', in Michael Young (ed.) *Knowledge and Control. New Directions for the Sociology of Education* London: Collier-Macmillan.

Bourdieu, Pierre (1998) *On Television* New York: The New Press.

Brewer, John and Porter, Roy (eds) (1993) *Consumption and the World of Goods* London: Routledge.

Broadcasters' Audience Research Board [BARB] (2007) 'Television Ownership in Private Domestic Households: 1956–2007' *Broadcasters' Audience Research Board,* online at: http://www.barb.co.uk/tvfacts (consulted on 6 June 2007).

Cammack, Paul (2006) 'Global Governance, State Agency and Competitiveness: The Political Economy of the Commission for Africa' *The British Journal of Politics and International Relations* 8 (3), pp. 331–350.

Carpenter, Humphrey (2002) *That Was Satire, That Was: The Satire Boom of the 1960s* London: Phoenix.
Carson, Rachel (2000; first published 1962) *Silent Spring* Harmondsworth: Penguin.
Castells, Manuel (2007) 'Communication, Power and Counter-Power in the Network Society' *International Journal of Communication* 1, pp. 238–266.
Cockett, Richard (1994) *Thinking the Unthinkable: Think-Tanks and the Economic Counter-Revolution, 1931–1983* London: HarperCollins.
Collini, Stefan (2002) 'Every fruit-juice drinker, nudist, sandal-wearer: Intellectuals as other people', in Helen Small (ed.) *The Public Intellectual* Oxford: Blackwell Publishing.
Corner, John and Pels, Dick (eds) (2003) *Media and the Restyling of Politics: Consumerism, Celebrity, Cynicism* London: Sage Publications.
Crisell, Andrew (1991) 'Filth, sedition and blasphemy: The rise and fall of television satire', in John Corner (ed.) *Popular Television in Britain* London: BFI.
Crouch, Colin (2004) *Post Democracy* Cambridge: Polity Press.
Darlow, Michael (2004) *Independents Struggle: The Programme Makers Who Took On The TV Establishment* London: Quartet Books.
Darlow, Michael (2004a) 'Behind the Goatees' *The Guardian* 3 September.
Darlow, Michael (2006) 'Hard Times – Some independent TV Producers Behave like Victorian Mill Owners' *The Guardian* 23 March.
Davis, Aeron (2002) *Public Relations Democracy: Politics, Public Relations and the Mass Media in Britain* Manchester: Manchester University Press.
De Zengotita, Thomas (2005) *Mediated: How the Media Shape Your World* London: Bloomsbury.
Debord, Guy (1996; first published 1967) *The Society of the Spectacle* New York: Zone Books.
Dinan, William and Miller, David (eds) (2007) *Thinker, Faker, Spinner, Spy: Corporate PR and the Assault on Democracy* London: Pluto Press.
Dovey, Jon (2000) *Freakshow: First Person Media and Factual Television* London: Pluto Press.
Dyke, Greg (2004) *Inside Story* London: HarperCollins.
Edwards, David and Cromwell, David (2006) *Guardians of Power – The Myth of The Liberal Media* London: Pluto Press.
Ellis, John (2000) 'Scheduling: The Last Creative Act in Television?' *Media, Culture and Society* 22 (1), pp. 25–38.
Energy Bulletin (2004) online at: http://www.energybulletin.net (consulted on 12 April 2008).
Evans, Peter (2005) 'Counterhegemonic globalization: Transnational social movements in the contemporary Global Political economy', in Thomas Janoski, Robert Alford, Alexander Hicks and Mildred A. Schwartz (eds) *The Handbook of Political Sociology: States, Civil Societies and Globalization* New York: Cambridge University Press.
Fisher, F. J. (1948) 'The Development of London As a Centre of Conspicuous Consumption in the Sixteenth and Seventeenth Centuries' *Transactions of the Royal Historical Society* 4th Series, 30, pp. 37–50.
Fuller, Graham (ed.) (1998) *Loach on Loach* London: Faber and Faber.
Gelpke, Basil and McCormack, Ray (2008) *A Crude Awakening* [DVD] London: Artificial Eye.

Gibney, Alex (2007) *Enron: The Smartest Guys in The Room* [DVD] London: Lions Gate.
Glasgow University Media Group (1976) *Bad News* London: Routledge and Kegan Paul.
Global Public Media (2001–) online at: http://www.globalpublicmedia.com (consulted on 12 April 2008).
Goddard, Peter (2004) 'World in action', in Glen Creeber (ed.) *Fifty Key Television Programmes* London: Arnold.
Goodin, Robert E. (1994) 'Selling Environmental Indulgences' *Kyklos* 47 (4) pp. 573–596.
Goodwin, Peter (1998) *Television Under the Tories: Broadcasting Policy 1979–1997* London: BFI Publishing.
Granovetter, Mark (1985) 'Economic Action and Social Structure: The Problem of Embeddedness' *American Journal of Sociology* 91, pp. 481–510.
Green TV (2006–) online at: http://www.green.tv (consulted on 10 April 2008).
Greene, Gregory (2004) *The End of Suburbia* [DVD] Ontario: The Electric Wallpaper Co.
Greenwald, Robert (2006) *Wal*Mart – The High Cost of Low Prices* [DVD] London: Tartan Video.
Guggenheim, Davis (2006) *An Inconvenient Truth* [DVD] London: Paramount.
Hari, Johann (2007) 'Yes the BBC is Biased – But to the Right' *The Independent* 9 April.
Harrison, Brian (1999) 'The Rise, Fall and Rise of Political Consensus in Britain since 1940' *History* 274, pp. 301–324.
Harvey, David (2005) *A Brief History of Neoliberalism* Oxford: Oxford University Press.
Harvie, David, Keir, Milburn, Trott, Ben and Watts, David (eds) *Shut Them Down! The G8, Gleneagles 2005 And The Movement of Movements* Leeds and Brooklyn: Dissent/Autonomedia.
Hay, Colin (2002) *Political Analysis* Basingstoke: Palgrave Macmillan.
Hay, Colin (2007) *Why We Hate Politics* Cambridge: Polity Press.
Heyck, Thomas W. (1998) 'Myths and Meanings of Intellectuals in Twentieth Century British National Identity' *Journal of British Studies* 37, pp. 192–221.
Hodkinson, Stuart (2005) 'Inside the Murky World of the UK's Make Poverty History Campaign' *Focus On The Global South*, 26 June (consulted 2 June 2007), online at: http://www.focusweb.org/content/view/810/27/.
Hodkinson, Stuart (2005a) 'Make the G8 History' *Red Pepper* July (consulted 2 June 2007) online at: http://www.redpepper.org.uk/global/x-jul05-hodkinson.htm.
Hodkinson, Stuart (2005b) 'G8 – Africa Nil' *Red Pepper* November (consulted on 2 June 2007), online at: http://www.redpepper.org.uk/global/x-jul05-hodkinson.htm.
Hopkins, Rob (2008) *The Transition Handbook: From Oil Dependency to Local Resilience* Totnes: Green Books.
Hubbard, Gill and Miller, David (eds) (2005) *Arguments Against G8* London: Pluto Press.
Isaacs, Jeremy (1989) *Storm Over Four: A Personal Account* London: Weidenfeld and Nicholson.
Isaacs, Jeremy (2007) *Look Me in the Eye – A Life in Television* London: Abacus.

Jansson, Andre (2002) 'The Mediatization of Consumption' *Journal of Consumer Culture* 2 (1), pp. 5–31.
Kampfner, John (2005) 'A very corporate loss of nerve' *New Statesman* 10 (October) (consulted 6 February 2007), online at: http://www.newstatesman.com/200510100006.
Karliner, J. (2001) 'A Brief History of Greenwash', *CorpWatch* 22 March (consulted on 10 May 2007), online at: http://www.corpwatch.org/article.php?id=243.
Kunstler, James Howard (2005) *The Long Emergency: Surviving the Converging Catastrophes of The Twenty-First Century* London: Atlantic Books.
Levi-Faur, David (2005) 'The Global Diffusion of Regulatory Capitalism' *Annals of the American Academy of Political and Social Science* 598, pp. 12–32.
Leys, Colin (2003) *Market Driven Politics: Neoliberal Democracy and the Public Interest* London: Verso Books.
Louw, Eric (2005) *The Media and Political Process* London: Sage Publications.
Mair, Peter (2006) 'Ruling the Void' *New Left Review* 42, November–December, pp. 25–51.
Marshall, George (2007) 'Why Rock Won't Save the Planet' *The Guardian* 5 July.
McCormick, John (1995) *The Global Environmental Movement* Chichester: John Wiley and Sons.
McLuhan, Marshall (2001; first published 1964) *Understanding Media: The Extensions of Man* London: Routledge.
Merchant, Carolyn (2005) *Radical Ecology: The Search for a Livable World* London: Routledge.
Mertes, Tom (ed.) (2003) *The Movement of Movements: A Reader* London: Verso Books.
Meyer, Thomas (2002) *Media Democracy: How the Media Colonize Politics* Cambridge: Polity Press.
Miller, David and Dinan, William (2008) *A Century of Spin – How Public Relations Became the Cutting Edge of Corporate Power* London: Pluto Press.
Monbiot, George (2006) 'The Denial Industry' in *Heat: How to Stop the Planet Burning* London: Allen Lane.
Morgan, Faith (2006) *The Power of Community: How Cuba Survived Peak Oil* [DVD] Yellow Springs, OH: The Community Solution.
Moseley, Rachel (2000) 'Makeover Takeover on British Television' *Screen* 41 (3), pp. 299–314.
Oborne, Peter (2007) *The Triumph of the Political Class* London: Simon and Schuster.
Office of Communications [Ofcom] (2006) *The Communications Market 2006* London: Ofcom.
Oliver, Michael and Pemberton, Hugh (2004) 'Learning and Change in 20th Century British Economic Policy' *Governance* 17 (3), pp. 415–441.
Palmer, Gareth (2003) *Discipline and Liberty: Television and Governance* Manchester: Manchester University Press.
Palmer, Gareth (2004) 'The new you: Class and transformation in lifestyle television', in Su Holmes and Deborah Jermyn (eds) *Understanding Reality Television* London: Routledge.
Paxman, Jeremy (2007) 'The James MacTaggart Memorial Lecture' *BBC Newsnight Blog* 24 August, online at: http://www.bbc.co.uk/blogs/newsnight/2007/08/the_james_mactaggart_memorial_lecture.html (consulted on 6 September 2007).

Payne, Anthony (2006) 'Blair, Brown and the Gleneagles Agenda: Making Poverty History, or Confronting the Global Politics of Unequal Development?' *International Affairs* 82 (5), pp. 917–935.

Plumb, John H. (1973) *The Commercialisation of Leisure in Eighteenth Century England* Reading: University of Reading.

Porteous, Tom (2008) *Britain in Africa* London: Zed Books.

Preston, Alison (2003) *Inside the Commissioners – the culture and practice of commissioning at UK broadcasters* Glasgow: The Research Centre for Television and Interactivity.

Quarmby, Katharine (2005) 'Why Oxfam Is Failing Africa' *New Statesman* 30 May (consulted 6 May 2007), online at: http://www.newstatesman.com/200505300004.

Rowell, Andrew (1996) *Green Backlash: Global Subversion of the Environment Movement* London: Routledge.

Sklair, Leslie (1997) 'Social Movements for Global Capitalism: The Transnational Capitalist Class in Action' *Review of International Political Economy* 4 (3), pp. 514–538.

Sklair, Leslie (2001) *The Transnational Capitalist Class* Oxford: Blackwell.

Snoddy, Raymond (2006) 'Bob Geldof: The Millionaire Media Player' *The Independent* 10 April.

Switzer, Jacqueline Vaughan (1997) *Green Backlash: The History and Politics of Environmental Opposition in the U.S.* London: Lynne Rienner Publishers.

Tunstall, Jeremy and Machin, David (1999) *The Anglo-American Media Connection* Oxford: Oxford University Press.

Veblen, Thorstein (1994; first published 1899) *The Theory of the Leisure Class* New York: Dover Publications.

Wackernagel, Mathis and Rees, William (1996) *Our Ecological Footprint: Reducing Human Impact on the Earth* Gabriola Island, BC: New Society.

Waltz, Mitzi (2005) *Alternative and Activist Media* Edinburgh: Edinburgh University Press.

West, Patrick (2004) *Conspicuous Compassion: Why Sometimes it Really is Cruel to be Kind* London: Civitas.

Williams, Paul D. (2005) *British Foreign Policy Under New Labour, 1997–2005* Basingstoke: Palgrave Macmillan.

Williams, Paul D. (2005a) 'Blair's Commission for Africa: Problems and Prospects for UK Policy' *The Political Quarterly* 76 (4), pp. 529–539.

Wilson, Edward O. (1992) *The Diversity of Life* Cambridge, Massachusetts: The Belknapp Press.

Wilson, Edward O. (1993) 'Is Humanity Suicidal?' *New York Times Magazine* 30 May, pp. 24–29.

Wilson, Edward O. (1997) *In Search of Nature* London: Penguin Books.

Woodward, John (1990) 'Day of the Reptile: Independent Production Between the 80s and the 90s', in Richard Paterson (ed.) *Organising for Change* London: BFI Publishing.

Wrigley, E. A. (1988) *Continuity, Chance and Change – The Character of the Industrial Revolution in England* Cambridge: Cambridge University Press.

Wrigley, E. A. (1992) 'Why poverty was inevitable in traditional societies', in John A. Hall and I. C. Jarvie (eds) *Transition to Modernity: Essays on Power, Wealth and Belief* Cambridge: Cambridge University Press.

Part II
Value, Hedonism, Critique

5
The Bohemian Habitus: New Social Theory and Political Consumerism

Sam Binkley

'Black Friday' is the name given to the first Friday after American Thanksgiving – the symbolic ending of the domestic serenity surrounding that occasion, and the launch of the frenzied holiday shopping season. But Black Friday has also recently become a day of reflection and protest directed against the culture of consumption itself, designated international 'Buy Nothing Day' by the anti-consumerist campaign group, Adbusters. In 2004, activities around Buy Nothing Day prompted a CNN interview with Adbusters founder and activist, Kalle Lasn, in which the following exchange took place:

> CNN: ... Kalle, I mean, Black Friday is like a tradition. People love to go out on this day and shop. We absolutely love it. Why do you want them to quit shopping?
> LASN: But think about it. After this very spiritual holiday of Thanksgiving, why is it that our culture is somehow then requiring us to go out the next day and max out on our credit cards... overconsumption is in some sense the mother of all our environmental problems.
> CNN: Oh, come on! Environmental problems?
> LASN: Yes, environmental problems.
> CNN: Oh, come on! Come on! If somebody wants to buy their kid an Elmo doll, what's the harm in that?
>
> (CNN transcripts, 2004)

The tone of this exchange is familiar to audiences of American television journalism in a post-FoxNews world, in which figures of the loony-left are routinely paraded for ridicule by a brutish television anchor, and where data-studded arguments evoking broad global pictures wither under the jeers of skeptical, everyday 'common sense'. But this encounter, I would argue, is interesting for other reasons, and serves here as a springboard into the central concerns of this chapter: the viability of political consumerism as a cultural and intellectual project, and the ambivalence or hostility with which it is greeted by some consumers.

Arguments for political consumerism have been offered by activists, scholars, students and everyday consumers themselves. In different ways, the claim has been made that mundane choices in the field of consumption might serve as instruments of ethical action and or have political impact (Conca *et al.*, 2002; Klein, 2000; Lasn, 1999; Micheletti, Follesdal and Stolle, 2003; *The ANNALS of the American Academy of Political and Social Science,* 2007; Stolle, 2005; *Cultural Studies,* 2008). It is hoped that consumers might be prompted to link their choices in the mall or grocery store to their broader global and societal consequences, and that social movements organised around these everyday actions of political consumerism might take the place of the grander strategies previously played out in more traditional political processes (Micheletti, 2003). While proponents of this new politics draw support from a variety of academic fields, the trend has been to pass over the work of traditional political theorists (whose approaches tend to center on the state and traditional forms of civic participation), in favour of macro-level explanations that are seen as better able to comprehend the changing configurations of politics and society on a global scale (Bauman, 1999; Beck *et al.*, 1997; Giddens, 1998; Walsh, 2004). Among such theorists, macro-level processes of social change are described in terms of the shifting location of politics itself: no longer enshrined in the traditional institutions of democratic governance, politics today is found in more spontaneous grassroots networks centered on personal, daily life concerns (Dalton and Wattenberg, 2002; Giddens, 1991; Rojek, 2001).

Indeed, recent mobilisations around consumption have found a fit with these sociological theories, drawing on notions of 'subpolitics' or 'life politics' described by Anthony Giddens, Zygmunt Bauman and most importantly Ulrich Beck, wherein personal practices in everyday life attain the significance of collective political action (Bauman, 1999, 2000; Beck *et al.*, 1994; Giddens, 1991; Holzer and Sørensen, 2003). Yet as Lasn's experience on CNN suggests, the extent to which an ethics of political consumption is influencing everyday lives and the practical logics of shopping encounters some obstacles, at least where it concerns such mundane objects as Elmo dolls and the like. The aim of this chapter is to locate these obstacles within the terrain of social theory: while theoretical invocations of subpolitics are doubtless promising, and the efforts of activists within the realm of consumption have already shown, and will doubtless continue to show, significant political traction, there remains much to be understood about the politicisation of everyday practices.

Against the backdrop of arguments for subpolitics and life politics gleaned from the sociology of Beck and Giddens, and drawing on Pierre Bourdieu's conception of 'habitus' as a pre-reflexive set of generative categories organising everyday practice, this chapter considers how the everyday character of mundane consumption practices, embodied in the pre-thought categories of the consumer's habitus, limits the viability of some anti-consumerist strategies as sub-political projects, particularly as they attempt to enlist wider constituencies (Bourdieu, 1977, 1984, 1997). Everyday consumption is defined by a particular entropy, or an inertia of the quotidian characterised by a logic of naturalness and common sense which tends to censor exhortations to ethical reflection and self-distancing, responding instead to the more immediate and practical logic of simply 'getting by'.

Yet rather than resign in despair at these limits, my aim here is to overcome them, to rethink political consumerism as a sphere of intervention by better understanding changes in the practical dynamics of everyday consumption. In the section on 'The bohemian habitus: a place for political consumerism?', I shall point the way towards further research into the changing configuration of the consumer habitus among influential segments of the middle classes of post-industrial societies (Harvey, 1992; Touraine, 1971). Drawing on recent analyses of the increasing aestheticisation of the economy developing from the growth of a largely urban, post-Fordist professional sector centered on the production of cultural goods, I shall argue that the increasing influence of inner city sub-cultural vanguard groups – or bohemians – promises to bring significant changes to the way people think and act as consumers, and thereby to the place of politics in daily habits of consumption (Entwistle and Wissinger, 2006; Florida, 2002b, *The Rise of the Creative Class*).

Life politics and subpolitics

Discussions of the politics of consumption have drawn from a range of sources in the field of social theory, to explain how the personal, mundane and seemingly inconsequential domain of consumption might assume the significance of politics. The arguments of Beck and Giddens on the rise of 'reflexive modernity', and what they respectively term 'subpolitics' and 'life politics', have in this regard served a special purpose. Together these theorists shed light on the proposed link between the personal realm of everyday life and new social movements centered on political consumerism.

The important points can be briefly summarised: Beck's comprehensive sociological account of what he terms 'reflexive modernity' describes a shift from a first-order modernity defined by the imperatives of progress, the domestication of nature, increased societal rationalisation and robust wealth generation to a second-order modernity in which previously concealed, unanticipated effects of these primary processes have emerged as urgent concerns in their own right (Beck, 1992). At an earlier moment of development driven by industrial modernisation, the imperative to tame nature and to satisfy fundamental human needs stood out as an enduring problem. Now, under reflexive modernity, it is the secondary effects generated by modern progress itself that warrant intervention. Beck has in mind a variety of environmental and health issues, but also a range of personal and cultural effects, from the erosion of community and personal isolation to the depletion of shared meanings (Beck, 1996). More precisely, these second-order effects of reflexive modernity involve the struggle of individuals to manage risks in their daily lives: once contained and minimised by the planning mechanisms of industrial modernity and the calculations of the welfare state, risks are today distributed multifariously throughout the social fabric. And they are left to the individual to negotiate through her own life plans, which include job training and retraining, the purchase of insurance policies, the maintenance of personal health and the like. In a 'risk society', everyday life is increasingly reflexive – examined and assessed by individuals themselves acting on their own, on themselves, without the support of the state or any collective body.

All of this has meant a specific shift in the location of politics: where previously struggles developed around those steering mechanisms by which the direction of modern advance was determined and its attendant risks were contained (particularly the welfare state, which set priorities for economic growth and wealth distribution but also supplied social safety nets to enforce risk reduction), today political processes take place within the spaces of those second-order consequences and risks once excluded from the realm of politics. Politics, in other words, has taken root in social and personal life, wherein modernity's unintended effects are experienced and negotiated on a daily basis (Beck *et al.*, 1997).

Moreover, at the centre of this shift is a process Beck describes as one of individualisation (Beck and Beck-Gernsheim, 2002). Individualisation entails the redefinition of core existential certainties – assured beliefs in community, shared purpose and meaning – from forms enshrined in meta-narratives of modern progress and shared institutions (such as trade unions, parties and civic organisations) to personal undertakings and objects of individual improvisation. With the bankruptcy of such meta-narratives and their legitimising discourses, we are left to rethink the basic assumptions and re-examine the daily effects of modern progress privately, in our own personal lives (David and Wilkinson, 2002). Previously taken-for-granted beliefs about the moral meanings underpinning modernisation – the expansion and rationalisation of industry, the growth of markets, the spread of administration and control mechanisms in daily life – are now viewed with suspicion as contingent events, measured for their consequences on social and personal existence. Reflexive modernity is perpetually assessed for the risks it incurs, and subjected to the rigours of a new kind of political scrutiny, or what Giddens calls a 'life politics' (Giddens, 1991).

Individualisation, however, does not necessarily entail atomisation. The improvisations of individuals in their daily lives can ultimately acquire a collective character as groups begin to form around the containment of risk and the replenishing of meaning. Taking up the newly politicised domain of personal experience in modern life wherein the unintended consequences of modernist development are felt the most intensely – areas such as ecology, personal health, community, consumption, well-being and so on – new political horizons open up which blur distinctions traditionally maintained between public and private life (Bennett, 2003; Halkier, 2001; Harvey, 1999; Knight and Greenberg, 2002). The case for anti-consumerism is often advanced against the backdrop of such a collapsing distinction: Michelle Micheletti, for example, has described the subpolitical as 'responsibility-taking by citizens in their everyday, individual-oriented life arena that cuts across the public and private spheres' (Micheletti, 2003, p. 29). Indeed, she goes on to offer the following example as an instance of subpolitics in action in the realm of consumption:

> Individuals begin by worrying about a private matter – wanting to provide a healthy meal for the family, work, a shorter day for personal health and family solitude, or buy new furniture for a barbecue planned on the patio – and soon find that their private issues and interests have a public side to them as well... Healthy food for one's family may mean finding where one can buy it, leading to a demand for organic foods and a movement for eco-labeled produce that takes a stance against genetically modified organisms, and finally

> in institutions that audit and label food products to ensure their environmental quality.
>
> (Micheletti, 2003, p. 35)

Micheletti's account illustrates a process central to political consumerist practice and theory: everyday concerns provide the spark which stimulates greater reflexivity or self-awareness, which in turn motivates a programme of political conduct. But central to this argument is an account of the increasingly self-conscious nature of daily life under the conditions of reflexive modernity – a point that is important for a theory of political consumerism, yet one which nonetheless runs up against some conflicting empirical realities.

It is Giddens who, in recognising this tension, has ventured furthest in exploring increasing levels of self-consciousness in reflexive modernity. For Giddens, reflexive modernity entails the extension of self-awareness into the most intimate domains of identity and selfhood. Individuals resolve existential dilemmas imposed by reflexive modernisation through a project of self-actualisation, at the centre of which is an increase in self-awareness, or 'reflexive self-monitoring', in all areas of life. Distinguishing life politics from more traditional forms of emancipatory politics, Giddens writes:

> Life politics concerns political issues which flow from processes of self-actualisation in post-traditional contexts, where globalizing influences intrude deeply into the reflexive project of the self, and conversely where processes of self-realisation influence global strategies.... Life politics, to repeat, is a politics of life decisions.
>
> (Giddens, 1991, p. 214f)

Giddens' thesis on life politics, like Beck's subpolitics, has fostered a fruitful research discourse on the proliferation of new social movements outside the categories defined by traditional politics, which derive from people's increasing self-awareness of personal experiences and daily practices (Bennett, 2003). Yet both assume a very high level of reflexivity in everyday life – degrees of self-awareness and self-consciousness, and a willingness to introduce highly reflexive ethical discourses into the most mundane aspects of one's daily practices. The question remains: To what extent can everyday consumption be defined by such high levels of self-awareness, or, conversely, to what extent do consumers possess, and practice, the ability to suppress self-awareness when it comes down to completing their daily consumption activities? This question, I contend, is best addressed through reflection on the role of the embodied logic of the habitus in consumption routines, and in the contrasting ways in which such forms of embodiment are conceived, on the one hand, as the object of reflexive self-awareness, and, on the other, as the pre-reflexive, unthought basis for everyday practice. To pose this question with some measure of theoretical clarity, we must turn to another important contribution from social theory, one provided by Pierre Bourdieu and his theorisation of the habitus as the structuring foundation for everyday practice.

Reflexivity, embodiment and the consumer habitus

An important element in Beck's and Giddens' theories of reflexive modernity concerns the role of the body, which emerges as an object of increasing awareness and scrutiny, and of cultivation and care. With the onset of reflexive modernity, the body, it is claimed, becomes 'denaturalised' – its givenness transformed into an object of choosing within the realm of human control. As Giddens describes it:

> The body used to be one aspect of nature, governed in a fundamental way by processes only marginally subject to human intervention. The body was a 'given', the often inconvenient and inadequate seat of the self. With the increasing invasion of the body by abstract systems all this becomes altered. The body, like the self, becomes a site of interaction, appropriation and reappropriation, linking reflexively organized processes and systematically ordered expert knowledge. The body itself has become emancipated – the condition for its reflexive restructuring.
>
> (Giddens, 1991, p. 218)

Giddens has in mind here the various regimes of bodily cultivation that have come to characterise contemporary lifestyles, which include exercise and health practices, dietary regimes and myriad other practices of self-monitoring and self-awareness centred on physical well-being. Such concerns provide the framework for the kinds of life politics and subpolitics already discussed, powerful elements of which appear in varieties of political consumerism ranging from green consumerism to environmentalism to alternative foodways (Lewis, 2006). Indeed, in many ways the body, as an object of reflexive self-awareness, has a significant role to play in the shaping of self-identity, and is mobilised as an important counterpoint to what are perceived to be the impersonal machinations of the mass market, distributors of risks to bodily health and existential well-being. The political force of this reappropriation of the body is captured in the tension Giddens describes between 'personalisation and commodification', in which the market presents specific challenges to the project of self-realisation:

> For the project of the self as such may become heavily commodified. Not just lifestyles but self-actualisation is packaged and distributed according to market criteria... Yet commodification does not carry the day unopposed on either an individual or a collective level. Even the most oppressed of individuals – perhaps in some ways particularly the most oppressed – react creatively and interpretively to processes of commodification which impinge on their lives.
>
> (Giddens, 1991, p. 198f)

One might include among the ranks of such oppressed individuals familiar figures within anti-consumerist discourses, from the French farming activist Jose Bové (who famously bulldozed a McDonalds restaurant in rural France) to the anti-McDonald's documentarian Morgan Spurlock (director of *Supersize Me*, in which Spurlock himself explores the immediate health effects of a fast food diet) – opponents of commodified food systems whose experience with the incursion of such abstract systems registers on the level of the body. The reflexivity of such groups is, of course, highly variable: as Jo Littler has pointed out, reflexivity in anti-consumerist politics varies from 'a relatively narcissistic form of reflexivity that acts to shore up a romantic anti-consumerist activist self' to one that expresses a more relational and dispersed process (Littler, 2003 p. 229). Yet in either case, it is an awareness of the body, its cultivation and maintenance as the object of a reflexive project of self-actualisation that serves as the touchstone of a political consumerist project. The body becomes, for those pursuing such a project of self-identity, an object-to-be-decommodified through sustained practices of reflexive self-awareness, coupled together with strategies of collective action (Binkley, 2008).

This account, however, does not fully take into consideration the ways in which people live their bodies in the performance of daily tasks, such as consumption. Everyday trips to the mall, to Starbucks, to Ikea; visits to retailers of unhealthy or environmentally damaging goods produced under conditions of exploitation within a global economy – these visits involve undertakings in which actors inhabit their bodies in ways that specifically exclude much reflective thinking. Bourdieu's theorisation of social practice provides insight on this by directing us to the unthought, pre-reflexive features of the body in a range of daily tasks and interactions; it offers a form of understanding which, in my view, significantly advances the project of anti-consumerism as a political enterprise (Bourdieu, 1977, 1984, 1992; Kauppi, 2000).

Bourdieu's theory of social practice centres on the role of the habitus of the specific actors, understood as a system of bodily dispositions in which social locations are internalised, naturalised and experienced as the common-sense articulation of things, yet incorporated as a transposable set of bodily logics, or a 'bodily hexis', serving as a generative set of principles for the structuring of everyday practices (Bourdieu, 1977, pp. 93f.). The habitus is, Bourdieu writes, 'a system of lasting, transposable dispositions which, integrating past experiences, functions at every moment as a matrix of perceptions, appreciations, and actions, and makes possible the achievement of infinitely diversified tasks, thanks to the analogical transfers of schemes' (Bourdieu, 1977, p. 82f). Distinguished from the reflexive body of life politics, the habitus is the place in which prior determinations imprinted on the individual in the course of life – determinations originating in inter-group relations at the structuring of society, and most likely in the individual's specific class location – are converted into naturally felt and taken-for-granted aspects of social existence and daily practice. 'The habitus makes coherence and necessity out of accident and contingency', writes Bourdieu (Bourdieu, 1977, p. 87). Indeed, habitus shapes ways of acting premised on specific ways of perceiving the world, even as it integrates practices of 'apperception', or managed tasks of non-comprehension, into its mode of daily conduct (Bourdieu, 1977, p. 86). The habitus expresses deeply engrained patterns of perception and apperception, ways of being conscious of specific things but also ways of remaining specifically unconscious of them.

In this way, the principle of the habitus seems to demand careful qualification of claims concerning self-awareness and the power (or obligation) of individuals to make specific choices about their bodily well-being. Where life politics is premised on new forms of awareness directed at the body, the habitus seems specifically structured around the suppression of such awareness as a condition of its operation, foreclosing the very distance one takes on oneself when one considers one's actions ethical. Bourdieu writes:

> Through the systematic 'choices' it makes among the places, events and people that might be frequented, the *habitus* tends to protect itself from crises and political challenges by providing itself with a

> milieu to which it is as pre-adapted as possible, that is, a relatively constant universe of situations tending to reinforce its dispositions by offering the market most favourable to its products. And once again it is the most paradoxical property of the *habitus*, the unchosen principle of all 'choices', that yields the solution to the paradox of the information needed in order to avoid information. The schemes of perception and appreciation of the *habitus* which are the basis of all the avoidance strategies are largely the product of a non-conscious, unwilled avoidance, whether it results automatically from the conditions of existence (for example, spatial segregation) or has been produced by a strategic intention (such as avoidance of 'bad company' or 'unsuitable books') originating from adults themselves formed in the same conditions.
>
> (Bourdieu, 1992, p. 61)

As such, the habitus, as the 'unchosen principle of all "choices"', deflects or suppresses views that threaten to unsettle it, to de-naturalise the naturalness by which it operates. While Bourdieu's critics on the left have questioned what they consider his fatalism on this point, his analysis provides insight on the wider puzzle confronting political consumerism (Bourdieu and Eagleton, 1999; Csordas, 1999; Holton, 2000; Wacquant, 1993). As proponents of political consumerism attempt to expand their ranks, their message can at times run up against the limits of the habitus: exhortations to consume more responsibly are resisted or ignored.

The bohemian habitus: a place for political consumerism?

Competing accounts of the viability of political consumerism as a practice incorporating a certain reflective distance taken by practitioners on their own practices have been offered here in order to clarify the likely limits on the current objectives of political consumerism. Yet the aim of this discussion is not to foreclose such objectives, but to consider more advantageous ways of realising them. Towards this end, I will offer some further comments on the developments in the nature of consumer practices that now and in the foreseeable future hold out opportunities for consumer activism. While more research is required before these points can be presented in anything like a synthetic manner, I offer them here in thumbnail form, as an affirmative conclusion to the points already made.

In a very general sense, in many areas of society, the consumer habitus is changing in significant ways. People are becoming less hostile to appeals to self-awareness in their everyday habits, at least among a small but influential portion of the population. This transformation is occurring along with a broader transformation in the economic organisation of capitalism from a Fordist model centered on the manufacture of mass produced goods to a post-Fordist mode centered on cultural, aesthetic and symbolic production (Harvey, 1992). Very briefly, contemporary forms of capitalist development have turned from a model of growth and accumulation centered on competition in the realm of industrial production (where techniques of shop-floor discipline and the rational planning on the managerial level assured strong, regular and voluminous output of serially produced goods) to one centred on the production of experiences, meanings, signs, knowledge and abstract values – to what has been termed the 'culturalization' of the economy (Du Gay and Pryke, 2002). Under these conditions, the occupational structure of Western economies has undergone a radical change: once dominated by workers and managers accustomed to hierarchical organisational schemes and dutiful, if repetitive work, today it involves a far more flexible workforce – designers, advertisers, marketers, communicators of various stripes, aestheticians, and 'cultural intermediaries' – who drive the economy (or large portions of it). Such a case is advanced persuasively by Scott Lash and John Urry in *Economies of Signs and Space*, wherein the authors argue that 'economic and symbolic processes are more than ever interlaced and inter-articulated . . . the economy is increasingly culturally inflected and . . . culture is more and more economically inflected' (Lash and Urry, 1994, p. 64).

Moreover, these changes, it is claimed, have cultivated a workforce endowed with the capacity to innovate aesthetically, to communicate in nuanced and expressive ways, and to mediate the world of appearances in new ways. This has been noted in several influential urban centres (New York, London, Berlin), where capital accumulation draws on innovations in the fields of research, media and expressive culture, and the work of 'cultural mediation' has become central to economic development (McRobbie, 1999; Lloyd, 2006; Zukin, 1982, 1995). Labour has been drawn from growing pockets of urban bohemians – artists, musicians, intellectuals, small business people, writers and others – whose numbers have been rising since the 1980s. Speaking of the American context, Richard Florida has described transformations in the workforces of such centres in terms of the 'rise of the creative class': individuals with aesthetic sensibilities, Florida argues, are seminal to economic growth in urban centres, where their talents feed the growing culture industries and their mere presence in certain neighbourhoods increases a city's cultural appeal, giving it an edge over other urban centres in an increasingly competitive global marketplace (Florida, 2002a, 2002b). Indeed, the economic contributions of bohemians have been judged to be of such significance as to justify their extensive monitoring, as evidenced in the development of the 'Bohemian Index', an inventory of writers, artists and performers compiled by Florida. For the purposes of the Index, a bohemian is described as 'someone who believes they cannot be defined by their job. Bohemians devote their lives to the pursuit of things other than money, but end up with a stable income. Bohemians believe that there is a class system in America, and believe themselves exempt' (Florida, 2005; Womack, 2004). Yet to understand who these workers are, how they go about the mundane tasks of consumption and what sort of specific habitus they embody, we must reflect briefly on the history and structural location of this group within core social dynamics of modern societies. This location, I will argue, can be linked to the legacy of bohemianism more generally.

Historically, bohemianism defines an aesthetic disposition which brings together a romantic investment in the authenticity and irreducible autonomy of aesthetic production as a practice of everyday life – one which operates against the perceived encroachments of a capitalist market in cultural goods patronised by bourgeois audiences (Bourdieu, 1993; Wilson, 2000). While traditionally such an antagonistic stance consigned the bohemian to the margins of economic life, today, bohemian sensibilities are well incorporated into market systems – part and parcel of economic 'culturalization'. The antagonism between the instrumental demands of economic growth and the expressive possibilities of leisure and consumption that worried social theorists like Daniel Bell (1976) (who fretted over the 'disjuncture' between the cultural and economic spheres) has been resolved, as the expressive logics of bohemia, with their penchant for authenticity and creativity in everyday life, have come to define more and more aspects of personal and commercial life.

Yet the oppositional quality and aesthetic self-awareness that shape the bohemian disposition run deeper than its professional incorporation. That disposition persists in many aspects of everyday life, perpetuated in a deeply internalised, intuitively felt way of getting by – what we might call a 'bohemian habitus'. Bohemianism reproduces an aristocratic disdain for bourgeois culture and its market products, enacting a stylised refusal of what it deems crass commercialism; it petitions on the side of personal integrity, authenticity and an embodied naturalness against the abstractions, calculations and impersonal machinations of mass society; yet at the same time it licenses the play of appearances and aesthetics in everyday life, and occasionally delights in the contradictions of its own position with regard to a market system it knows it cannot resist (and, indeed, upon which it remains dependent). Moreover, it is the bohemian's specific predilection for aesthetic manipulation – a delight taken in the stylisation of appearances and the collapsing of boundaries separating artistry and daily life – that situates bohemian antagonism to the market. The bohemian looks with scorn upon the regimented production of serially produced, identical goods, which suppresses the expressive capacities and the everyday artistry to which the bohemian remains committed.

For this reason, the bohemian habitus is at once antagonistic and reflexive: the integrity of aesthetic production demands a constant self-monitoring and an opposition to the instrumental rationality of the market system. This 'aestheticization of everyday life', as Mike Featherstone (1990) has termed it, demands an awareness of the self as an object of aesthetic consumption and production. With the increasing culturalisation of economic and social life, such sensibilities, it can be argued, have become significantly generalised, such that the oppositional stance of this new bohemianism has become a general feature of many consumer markets and practices. This is evidenced in the growing popularity of personalised products, appeals to the expressive uniqueness and alleged creative agency of individual consumers merely through the act of purchase, and increasing flirtations with the ethical dimensions of product choice, evidenced in Gap's 'Red' campaign, Starbuck's emphasis on economic justice in the purchase of coffee beans, and the general 'greening' of consumer culture. While much of this falls radically short of a sustained political strategy (such a 'greening' of consumption, it is often pointed out, amounts to little more than a 'greenwashing' of mass market practices), what is important here is the oppositional logic of the habitus it betrays, and the potential that such a shared habitus holds for more radical appeals to integrate reflexive awareness into the daily practices of shopping and spending. Such an oppositional logic can be apprehended structurally and historically in the location held by bohemians, as practitioners of a reflexive aesthetics of daily life.

Moreover, reflexivity and opposition among bohemians, as any stroller through an urban enclave registering high on the Bohemian Index can attest, is not a consciously held position, but an embodied logic, embedded in an everyday mode of embodiment. The body is here inhabited as the seat not only of a replenishing source of expressive authenticity in everyday practice, but also as the embodied negation of the commodity form itself: against the seriality and standardisation of mass market products, the bohemian body exudes expressive authenticity and opposition, even as this opposition is becoming increasingly central to the production and circulation of goods for those same markets. While critics have convincingly pointed out the extent to which this opposition depends on an ever more implausible differentiation between a bohemian space of expressive authenticity and a commodified space of mass consumption, it remains the case that the belief in this distinction is sustained in the minds, bodies and practices of bohemians (and those inflected with the logic of their practice), animating their conduct and opening them to perhaps more radical critical discourses on consumption as an ethically consequential mode of conduct (Frank, 1997; Heath and Potter, 2004).

In other words, the bohemian habitus mobilises the body against the market in its own way, and in doing so overcomes the foreclosure of reflexive politics evidenced by many other consumer groups. Where the commodifying logic of the mass market is marked by antiseptic, abstract, rationalist and hygienic (and hence bodiless) textures, the bohemian disposition cultivates a sense of authentic corporeality in its own daily practices. This taste can be witnessed in any of thousands of bohemian coffee shops which signal their independence from the sterility, homogeneity and institutional sensibility of large franchises by cultivating a sense of authenticity, warmth and 'funkiness' (Thompson and Arsel, 2004). It is palpably expressed in the bohemian somatic practice of tattooing and body modification, in countercultural youth styles, much of which seems to cut against the cold seriality of the commodity form. Indeed, this embodied opposition, which marshals the body's authenticity and sensuality against the coldness and instrumentality of commodity culture, incorporates reflexivity into the practice of everyday consumerism.

One might argue, then, that as these cultural vanguards extend their influence more generally, more people are likely to be inflected with a specifically bohemian sense of aesthetics in daily life, and to develop an implicit opposition to a market system deemed lacking in corporeal pleasures and expressive embodiment. Moreover, it seems possible that such conditions would mean that consumers became more responsive to appeals for ethical reflection on their everyday consumer practices, thus making them more aware of their environmental impacts, consequences for health and implications for global economic inequality. The reflexive self-awareness that is part and parcel of subpolitics and life politics moves from being an explicit reflexive discourse to an internalised mode of daily practice – an embodied characteristic of the expressive body of the bohemian habitus itself, thus permitting an openness to self-reflection suppressed in many other walks of life.

Concluding reflections

In the preceding pages I have offered what I hope is a fairly convincing (though undoubtedly, for some, an unnecessarily lofty) account of some of the problems and opportunities facing the project of political consumerism. While it is widely held that political consumerism opens new avenues of political participation and contest, it does so by politicising a dimension of everyday conduct that poses more resistance to this politicisation than many consumption activists and scholars typically recognise. My comments on the reflexivity of the consumer habitus are intended as a cautionary note to readers whose inclinations have led them to a book that takes seriously the promise of this new political opening. At the same time, I have indicated that the changing field of consumption, aligned with broad economic and societal transformations concentrated in urban areas, yet apparent more generally in a range of sectors and populations, holds the promise for increasing receptivity among consumers to efforts to instill ethical reflexivity in their habits as consumers. Urban bohemian vanguards, whose distinctively reflexive sensibilities as consumers are disproportionately influential to their numbers, provide something of a template for more mainstream groups. I would like to offer in these closing remarks a reflection on the relevance of this argument to the wider aims of this volume, and to some more general challenges that face political consumerism as a social movement.

First with regard to the presupposition of this volume and the place occupied by this chapter within this wider polemic. It seems to me that, while my position is ostensibly critical, a certain complementarity is reached. While this chapter sets out to limit the claims of political consumerism, this effort concludes with a model of ethical consumerism that dovetails well with the 'alternative hedonism' that is proposed in this volume. The editors write in the volume's introduction: 'A counter-consumerist ethic and politics should therefore appeal not only to altruistic compassion and environmental concern, but also to the more self-regarding gratifications of consuming differently. It should develop and communicate a new erotics of consumption or hedonist "imaginary".' In that sense, I think we are on the same track. Such an erotics of consumption is certainly appropriate to the modes of embodied reflexivity expressed in the bohemian habitus. Indeed, the cautionary tone expressed in this chapter against robust and explicit exhortations to political consumption – exhortations, perhaps, by activists or scholars operating in a much more dogmatic mode than those discussed in this volume – is perhaps better saved for those with a more diminished vision of the aesthetic properties of the good life. Alternative hedonism seems to anticipate and incorporate many of the assertions put forward here, particularly in its call for a 'hedonist imaginary' – a task perhaps best tackled in the aesthetic domain to which urban bohemians are particularly well adapted.

Yet while the willingness to depart from the didacticism, dogmatism and self-consciousness of much conventional social movement discourse, and to invest the sensual realm of aesthetics and hedonism, might be a welcome departure for a politics of consumption and might take important steps in avoiding the stubborn refusals of the prereflexive habitus discussed here, it nonetheless opens itself up to the other problems which both the present article and the general thesis of alternative hedonism have yet to address satisfactorily. These concern the fundamental requirement that political consumerism, to be effective, necessarily entails the general availability of accurate information on the wider social, economic and environmental impact of consumption decisions. For political consumers to develop and practice habits which are only in a very general and ultimately aesthetic way oriented to wider political objectives runs the real danger that these decisions could become misdirected and fail to effect real political change.

This problem seems to run to the heart of claims for political consumerism as a social movement. Such claims rightly acknowledge the uniqueness of political consumerism in reaching new constituencies and eliciting political participation from wider groups than traditional social movements, by concentrating on non-traditional outlets for political participation, embedded in mundane, extra-institutional contexts, and embracing dispersed, non-hierarchical organisational forms. Yet such institutional and hierarchical forms served an important purpose in traditional social movements in orienting collective action towards strategic goals. Without such coordination, and without the dissemination of information and directives to its memberships, political consumerism runs the risk of remaining ineffectual and of missing key strategic targets, in spite of the enthusiasm of its participants.

Michele Micheletti has described in great detail the ways in which political consumerist efforts are often wide of their mark, and the need for accurate auditing mechanisms in political consumerist practice (Micheletti, 2003, pp. 9–11). Where snap decisions between various products undertaken in the midst of mundane shopping activities are guided only by rumour, by hunches or aesthetic inclinations, the likelihood is tenuous that they will successfully apply just the correct amount of pressure to the most economically and politically sensitive spots in the consumer market place. What is required, Micheletti argues, are accurate auditing mechanisms capable of informing and directing ethical consumer choices, strategists capable of devising plans and operations with specific goals in mind, and efficient and transparent means of communicating these plans to willing consumers (p. 128). In short, in addition to aesthetics, what is required is education. For this purpose, a range of labeling schemes has traditionally served to orient consumers, and now there are several auditing agencies and certification schemes, from fair trade organisations to the Clean Clothes Campaign. In other words, political consumerism, like the traditional social movements its advocates counterpoise it to, still runs up against some familiar difficulties: how to direct the activities of spontaneously acting participants in order that meaningful and pragmatic change might occur. The largely aesthetic and hedonistic solutions proposed in this chapter, and in this volume more generally, must account for the role of such distinctly informational, and at times dogmatic and instructional content, if political consumerism is to translate into an effective strategy for the monitoring and regulation of the global economy. This problem is doubly important when one considers the slipperiness of politics in the phantasmagoric world of contemporary marketing. Strategies such as greenwashing, and now bluewashing and sweatwashing, make the politics of aesthetic decision making highly problematic (p. 163). What is required, then, is an integration of aesthetic and informational content, in such a way that the hedonistic dynamics of consumption retain their persuasive appeal. Labeling schemes must partake in the aesthetics of hedonistic consumption without appearing dogmatic and didactic. And what is required is a sense of identification and trust with the auditing sources which relate information to consumers, and with the planning authorities – activists and organisers – whose presence must be integrated with this general aesthetics. Trust, a resource that is in radically short supply in late modern contexts, must be maintained by a leadership celebrated for its aesthetic, and not just moral or political stature.

References

Bauman, Zygmunt (1999) *In Search of Politics* Cambridge: Polity Press.
Bauman, Zygmunt (2000) *Liquid Modernity* Cambridge: Polity Press.
Beck, Ulrich (1992) *Risk Society* Thousand Oaks: Sage.
Beck, Ulrich (1996) 'World Risk Society as Cosmopolitan Society? Ecological Questions in a Framework of Manufactured Uncertainties', *Theory, Culture and Society* 13(4), pp. 1–32.
Beck, Ulrich and Elisabeth Beck-Gernsheim (2002) *Individualization: Institutionalized Individualism and its Social and Political Consequences* London: Sage.
Beck, Ulrich, Anthony Giddens and Scott Lash (1994) *Reflexive Modernization* Cambridge: Polity Press.
Beck, Ulrich, Anthony Giddens and Scott Lash (1997) *The Reinvention of Politics: Rethinking Modernity in the Global Social Order* Oxford: Polity Press.
Bell, Daniel (1976) *The Cultural Contradictions of Capitalism* New York: Basic Books.
Bennett, Lance (2003) 'Branded Political Communication: Lifestyle Politics, Logo Campaigns, and the Rise of Global Citizenship', in Michele Micheletti, Andreas Follesdal, and Dietlind Stolle (eds) *Politics, Products, and Markets: Exploring Political Consumerism Past and Present* New Brunswick: Transaction Publishers, pp. 101–125.
Binkley, Sam (2008) 'Liquid Consumption: Anti-Consumerism and the Fetishized De-Fetishization of Commodities', *Cultural Studies* 22, 5–6.
Bourdieu, Pierre (1977) *Outline of a Theory of Practice* Cambridge: Cambridge University Press.
Bourdieu, Pierre (1984) *Distinction* London: Routledge.
Bourdieu, Pierre and Wacquant, Loic (1992) *An Invitation to Reflexive Sociology* Chicago: Chicago University Press.
Bourdieu, Pierre (1993) *The Field of Cultural Production* New York: Columbia University Press.
Bourdieu, Pierre and Terry Eagleton (1999) 'Doxa and Common Life: An Interview', in Slavoj Zizek (ed.) *Mapping Ideology* London: Verso.
CNN Transcripts (2004) Anderson Cooper 360 Degrees: 'Will Ukraine's Disputed Election Lead to Civil War? Is Shopping for Holidays Overrated?', Aired November 26, 2004 at 19:00 http://transcripts.cnn.com/TRANSCRIPTS/0411/26/acd.01.html.
Conca, Ken, Michael Maniates and Thomas Princen (eds) (2002) *Confronting Consumption* London: MIT Press.
Csordas, Thomas J. (1999) 'Embodiment and Cultural Phenomenology', in Gail Weiss and Honi Forn Habor (eds) *Perspectives on Embodiment* London: Routledge.
Cultural Studies (2008) Special issue on Political Consumerism (editors: Sam Binkley and Jo Littler) 22, 5–6.
Dalton, Russell and Martin Wattenberg (eds) (2002) *Parties Without Partisans. Political Change in Advanced Industrial Democracies* Oxford: Oxford University Press.
David, Mathew and Ian Wilkinson (2002) 'Critical Theory of Society or Self-Critical Society?' *Critical Horizons* 3(1), pp. 131–158.

Du Gay, Paul and Michael Pryke (2002) 'Cultural Economy: An Introduction', in Du Gay and Pryke (eds) *Cultural Economy* London: Sage.

Entwistle, Joanne and Elizabeth Wissinger (2006) 'Keeping up Appearances: Aesthetic Labour in the Fashion Modelling Industries of London and New York', *The Sociological Review* 54(4), pp. 774–794.

Featherstone, Mike (1990) 'The Aestheticisation of Everyday Life', in Mike Featherstone, *Consumer Culture and Postmodernism* London: Sage.

Florida, Richard (2002a) 'Bohemia and Economic Geography', *Journal of Economic Geography* 2, pp. 55–71.

Florida, Richard (2002b) *The Rise of the Creative Class. And How It's Transforming Work, Leisure and Everyday Life* New York: Basic Books.

Florida, Richard (2005) 'Bohemian Rhapsody', *New York Times*, July 31, p. 11.

Frank, Thomas (1997) *Conquest of Cool* Chicago: Chicago University Press.

Giddens, Anthony (1991) *Modernity and self-identity* Cambridge, UK: Polity.

Giddens, Anthony (1998) *The Third Way* Cambridge, UK: Polity.

Halkier, Bente (2001) 'Consuming Ambivalences: Consumer Handling of Environmentally Related Risks in Food', *Journal of Consumer Culture* 1(2), pp. 205–224.

Harvey, David (1992) *The Condition of Postmodernity* Cambridge: Blackwell.

Harvey, David (1999) 'Considerations on the Environment of Justice', in N. Low (ed.) *Global Ethics and Environment* New York: Routledge, pp. 109–130.

Heath, Joseph and Andrew Potter (2004) *Nation of Rebels: Why Counterculture Became Consumer Culture* New York: Collins.

Holton, Robert (2000) 'Bourdieu and Common Sense', in Nicholas Brown and Imre Szeman (eds) *Pierre Bourdieu: Fieldwork in Culture* Lanham, MD: Rowman and Littlefield, pp. 87–99.

Holzer, Boris and Mads P. Sørensen (2003) 'Rethinking Subpolitics: Beyond the "Iron Cage" of Modern Politics?' *Theory, Culture and Society* 20(2), pp. 79–102.

Kauppi, Niilo (2000) *The Politics of Embodiment: Habitus, Power in Pierre Bourdieu's Theory* Frankfurt am Main: Peter Lane.

Klein, Naomi (2000) *No Logo: Taking Aim at the Brand Bullies* Toronto: Knopf.

Knight, Graham and Josh Greenberg (2002) 'Promotionalism and Subpolitics: Nike and Its Labor Critics', *Management Communication Quarterly* 15(4), pp. 541–570.

Lash, Scott and John Urry (1994) *Economies of Signs and Space* London: Sage.

Lasn, Kalle (1999) *Culture Jam: How to Reverse America's Suicidal Consumer Binge – And Why We Must* New York: Quill, Harper Collins Publishers.

Lewis, Tania (2006) 'DIY selves? Reflexivity and Habitus in Young People's use of the Internet for Health Information', *European Journal of Cultural Studies* 9(4), pp. 461–479.

Littler, Jo (2003) 'Beyond the Boycott: Anti-Consumerism, Cultural Change and the Limits of Reflexivity', *Cultural Studies* 19(2), pp. 227–252.

Lloyd, Richard (2006) *Neo-Bohemia: Art and Commerce in the Postindustrial City* New York: Routledge.

McRobbie, Angela (1999) *In the Culture Society: Art Fashion and Popular Music* London: Routledge.

Micheletti, Michele, Andreas Follesdal and Dietlind Stolle (eds) (2003) *Politics, Products, and Markets: Exploring Political Consumerism Past and Present* New Brunswick: Transaction Publishers.

Micheletti, Michele (2003) *Political Virtue and Shopping: Individuals, Consumerism, and Collective Action* New York: Palgrave.
Rojek, Chris (2001) 'Leisure and Life Politics', *Leisure Sciences*, 23, pp. 115–125.
Stolle, Dietlind (2005) 'Politics in the Supermarket: Political Consumerism as a Form of Political Participation', *International Political Science Review* 26(3), pp. 245–269.
The ANNALS of the American Academy of Political and Social Science (2007) Special Issue: The Politics of Consumption/The Consumption of Politics (editors: Dhavan V. Shah, Douglas M. McLeod, Lewis Friedland and Michelle R. Nelson) May, 611(1).
Thompson, Craig and Zeynep Arsel (2004) 'The Starbucks Brandscape and Consumers' (Anticorporate) Experiences of Glocalization', *Journal of Consumer Research* 31 (3), pp. 631–642.
Touraine, Alain (1971) *The Post-Industrial Society* New York: Random House.
Wacquant, Loïc (1993) 'From Ideology to Symbolic Violence: Culture, Class, and Consciousness in Marx and Bourdieu', *International Journal of Contemporary Sociology* 30(2), pp. 125–142.
Walsh, Mary (2004) 'Political Theory, Sociology and the Re-Configuration of Politics', Paper presented at the Political Theory Stream, British Political Studies Association, 54th Annual Conference, 6–8 April 2004, University of Lincoln, England.
Wilson, Elizabeth (2000) *Bohemians: The Glamorous Outcasts* New Brunswick: Rutgers University Press.
Womack, Andrew (2004) 'The Bohemian Index: Interview with Dorothy Gambrel', *The Morning News*, February 9, (www.themorningnews.org/archives/new_york_new_york/the_bohemian_index.php).
Zukin, Sharon (1982) *Loft Living: Culture and Capital in Urban Change* Baltimore: Johns Hopkins University Press.
Zukin, Sharon (1995) *The Cultures of Cities* London: Blackwell.

6
Sustainable Hedonism: The Pleasures of Living within Environmental Limits

Marius de Geus

A main question today is how we can all live comfortably and well without exceeding ecological limits. Recent warnings such as the 2007 IPCC and Stern reports concerning the depletion of resources, vastly diminishing biodiversity, global warming and rising sea levels suggest that in the long run modern materialistic lifestyles are simply untenable. How, then, can the citizens of 'the West' today be persuaded that it is worthwhile to reject contemporary conceptions of the 'good life' and to opt for a joyful and creative 'art of living' based on less materialistic pleasures, more ecologically responsible behaviour, and a different structure of consumption?

The dilemmas posed by ecological constraints have renewed the interest in classic philosophical questions such as 'What is the good life?' and 'How to live responsibly?' We must ask, too, what exactly we mean by 'the art of living' in an age in which we face the risks of irreversible damage to the environment. I here discuss early and more recent debates on 'the good life', and explore philosophical visions and perspectives on the 'art of living' (*ars vitae*), in which alternative ideas of pleasure, enjoyment and happiness play a crucial role.

I begin by analysing two dominant conceptions of the 'art of living', the moralistic and the hedonistic, before considering the longer term consequences for humanity and the environment of today's predominantly materialistic and hedonistic approach to the 'art of living'. A subsequent section explores the more obvious dangers and disadvantages of the moralistic conception. I turn lastly to questions concerning the fundamental principles of a sustainable hedonism and green 'art of living', and consider the various aspects and dimensions of this alternative, before offering some general conclusions.

The two dominant approaches to the 'art of living'

In the last decades there has been extensive interest in the philosophy of the 'art of living'. Prominent social thinkers and philosophers such as Hannah Arendt, Erich Fromm, Ivan Illich, Alexander Nehamas, Martha Nussbaum, John Rawls and Amartya Sen have contributed to the study of how to live wisely, justly and happily. Of course, the 'art of living' has been a central issue in the works of classical philosophers such as Solon, Aristotle, Plato, Epicurus and Cicero, and has also been discussed by such noted thinkers as Hobbes, Locke, Montaigne, Rousseau, Kant, Schopenhauer, Nietzsche, J. S. Mill, Thoreau and Emerson.

It seems, indeed, as if modern social and political philosophy is nowadays returning to its historical roots, by increasingly focusing on traditional themes such as 'how to live in a meaningful and satisfying way', 'how to make the right choices in life', 'how to find the true essence of Being' and 'how to cope with the frailty and unpredictability of human affairs' (Nehamas, 2000, pp. 6ff, 10; Nussbaum and Sen, 1993).

At the same time, however, there exists a real and serious dilemma in both classic, pre-modern, and modern writings on the philosophy of the 'art of living'. This is a dilemma closely linked to the much debated concept of 'the good life', and can be summed up in the following question: Is 'the good life' fundamentally related to living well, in a just, ethical and responsible way, or does it primarily concern the satisfaction of needs, the pursuit of happiness and the attainment of pleasure? (Veenhoven, 2002, p. 357ff). In other words, should the 'art of living' be formulated in terms of self-control, moderation and social responsibility, or in terms of personal happiness and hedonism? (Dohmen, 2002, pp. 14ff, 16ff; also Veenhoven, 2002, pp. 357ff, 359ff).

In the moralistic understanding, the 'art of living' is the general disposition towards a life of high ethical standards. The individual conforms to the commonly acknowledged behavioural rules and norms ensuring virtue and excellence of character. In the hedonistic understanding, the 'art of living' is the general disposition to live a life of pleasure (*hèdonè* literally means the enjoyment of pleasure). The individual gives priority to comfort, elegance and luxurious gratification.

In the history of political philosophy the moralistic thinkers, who include Solon, Plato, Aristotle, Rousseau, Thoreau, and, more recently, Erich Fromm, Ivan Illich, Martha Nussbaum and Amartya Sen, argue for restraints on production and consumption. 'The good life', they insist, is incompatible with excess, and abundance does not make for greater happiness or satisfaction. On this moralistic view, individual and social well-being is attained not in the quest for wealth and pleasure, but by relinquishing material satisfactions and curbing human wants. Indeed, the 'art of life' is about reducing needs and implementing a relatively frugal lifestyle. This approach consistently emphasises the advantages of a simpler lifestyle and the importance of caring for the natural environment: contact with nature is considered to be of great importance for a well-spent life (de Geus, 1999).

Aristotle is a prime example of the moralistic approach to the 'art of living'. In his *Ethics*, an eloquent defense of civic virtues, he develops an intriguing theory of justice, moderation and human happiness. For Aristotle, excellence of living and character is dependent on striking a balance between two extremes and harmonising pleasure with reason (Aristotle, 1975, p. 68ff). He notes that 'moral excellence is a mean between two forms of badness, one of excess and the other of defect, and is so described because it aims at hitting the mean point in feelings and action' (Aristotle, 1975, p. 73).

In his view, abundance, luxury and an opulent lifestyle are not the prerequisites for 'the good life': on the contrary, they often obstruct human felicity. Moderate possessions are sufficient to live autonomously and enjoy a life of happiness and bliss (Aristotle, 1975, p. 308). The central role of the Greek city state, he thought, was to stimulate civic virtues and in particular to foster good habits among the citizenry by means of education and legislation. The Aristotelian 'art of living' is, then, in essence, a matter of prudential learning. It is knowing how 'to set measures' to one's behaviour, to limit one's ambitions and desires, to exercise self-control and to behave in a dignified and socially responsible manner.

In contrast, the hedonistic philosophers – for example, Democritus, Leucippus, but also liberal political theorists such as Hobbes and Locke (see below), and Pascal in his *Pensées* – argue that the 'art of living' and human happiness are achieved by satisfying what are, in principle, unlimited desires and leading a comfortable or even luxurious and affluent lifestyle (Hobbes, 1974, p. 160ff; Locke, 1965, p. 342ff; Pascal, 1995, pp. 60–82). In their analysis, an increase in the production and consumption of goods and services is a necessary condition of living 'the good life'. The advantages of a comfortable lifestyle are here accentuated and nature is generally viewed as an instrument to satisfy the ever-increasing desires of humankind.

The liberal philosophers Thomas Hobbes (1588–1679) and John Locke (1632–1704) are interesting exponents of the hedonistic line of thinking. For instance, in *Leviathan*, Hobbes presents humans as self-moving machines whose actions are determined by either aversions or appetites. In his analysis, the general disposition of humans is to aspire to felicity, which he defines dynamically in terms of incessant desires:

> Felicity is a continuall progress of the desire, from one object to another; the attaining of the former being still but the way to the latter. The cause whereof is, that the object of man's desire, is not to enjoy once only, and for one instant of time; but to assure for ever the way of his desire. And therefore the voluntary actions of all men, tend, not only to the procuring, but also to the assuring of a contented life.
>
> (Hobbes, 1974, p. 161)

The 'art of living', according to Hobbes, consists in the ability to outwit competitors in the daily struggle for survival between individuals that takes place at all levels in society. Hobbes contends that in a liberal (market) society humans are essentially motivated by self-interest and unlimited appetites, and should not restrain their wants and desires, if they wish to live comfortably and well. His conception of the 'art of living' is thus based on 'possessive individualism'; the constant search for and acquisition of goods is an inextricable aspect of a materialist philosophy that provides the basis of modern Western consumer society.

Analysing Locke's *Second Treatise*, one finds that happiness and 'the good life' are similarly, if more implicitly, defined in terms of material gratification and property rights. Such happiness appears strongly dependent on achieving the highest possible level of consumption. Once again, then, 'the good life' appears to be equated with material progress, the acquisition of property and the expansion of consumer choice. Status and well-being are primarily measured in terms of property rights and consumer goods. As Rousseau was one of the first to point out, the result is a society dedicated to the expansion of needs and the promotion of ever more insatiable desires (Rousseau, 1997).

But the pursuit of a high-consumption, materialist lifestyle brings with it far-reaching and highly detrimental environmental consequences – as we are now experiencing in the form of global warming, climate change, depletion of natural resources and loss of biodiversity. The core problem is that in modern society the materialistic and hedonistic 'art of living' has resulted in ecologically 'vicious' forms of consumption, in part because they are in themselves excessive, in part because of their relative insulation from their environmental impacts.

The consequences for humanity, environment and society

One of the more worrying contemporary consequences is that the general belief of the hedonist outlook that pleasure is the main good has now become the conviction that the endless pursuit of material possessions is a collective right. In our era people feel that they are entitled to intense and novel experiences and to full and immediate satisfaction of their needs and wants. Contemporary hedonism is no longer the privilege of a wealthy elite but an expectation of the mass of Western people – who are usually today in a position to acquire the latest LCD television sets, notebooks and DVD recorders. Some commentators even hold that consuming is not only a 'right', but has now effectively become a 'moral and social obligation'. Buying and consuming in the pursuit of personal happiness and pleasure and in the hope of excitement are often seen as morally acceptable because they secure the economy's stability and enhance prosperity.

It is also noteworthy that modern consumer society's hedonism has shifted attention from the satisfaction of basic needs and wants to an unprecedented longing for luxurious goods and services. Provision for amusement is also nowadays highly commercialised, with the tourist and entertainment industries offering multi-functional 'leisure centres', recreation and amusement parks (Disneyland, Euro Disney and so on), ever more thrilling roller coaster rides, meticulously planned holiday trips and fully organised 'wilderness' adventures – as well as a plethora of animated virtual reality shows and computer games.

Modern hedonism has thus been co-opted by an extensive and sophisticated culture industry and ideology dedicated to providing agreeable distractions for young and old. Supposedly neutral in its means, ends and goals, in reality this development promotes the core values of materialism, hedonism and the moral right to instant gratification. It thus promotes a normative cultural ethic whose consumerism helps to sustain our contemporary neo-liberal growth economy.

Even more important is the fact that modern hedonism resists any closure of the gap between consumer desires and their satisfaction. This is the logical implication of the influential and compelling theory of consumerism elaborated by Colin Campbell in his *The Romantic Ethic and the Spirit of Modern Consumerism*. According to Campbell, today's excessive consuming will often entail dissatisfaction, since most consumers (even if not all) are trapped in the so-called cycle of 'desire-acquisition-use-disillusionment-renewed desire' (Campbell, 1989, p. 90). More specifically, this mechanism, whereby wants are permanently renewed and happiness reduced to a continual escalation of desire, is, as Hobbes claimed, constantly stimulating the restless and insatiable modes of consumption which are so characteristic of the prevailing materialistic and hedonistic approach to the 'art of living'.

Dangers and disadvantages of the moralistic approach

As we have seen, the moralistic line of thinking about the 'art of living' rejects any association of 'the good life' with excess: abundance, it argues, creates neither happiness nor satisfaction. Instead, individual and societal happiness depends on restraint, frugality and obedience to established norms of conduct. But such a perspective on the 'art of living' is clearly not without its dangers and disadvantages, and it is to these that I now turn.

A common countering argument maintains that in a liberal democratic state citizens are autonomous and have the right to live according to their own views of 'the good life' and on the basis of their own value systems, preferences and consumer desires. This means that the state must be neutral in respect of morals and norms and must in principle leave consumption patterns, lifestyle choices and personal interpretations of the 'art of living' to individual citizens themselves. Since consumption belongs to the domain of the private individual, the state should in principle refrain from interference in it (Dobson, 2003, pp. 158–173; Wissenburg, 1998, p. 43ff). Given the dominance of this viewpoint in an ever more individualistic and materialist-hedonist culture, support for measures restricting consumption is meagre. Freedom of choice and action will generally be thought more important than an unpolluted environment.

The increasing faith of Western liberal democracies in the market mechanism has also told against support for government intervention in the area of consumption, and discouraged any attempts to revise views about the 'art of living' and the form of the 'good life'. Nor are the citizens of the market economy currently inclined to subscribe to any determinate green vision of how to live.

Moreover, in modern society, government and producers are often much more worried about the economic risks of free market capitalism than about its possible ecological dangers. As Ulrich Beck has argued in his *Risk Society*, governments have shown themselves more ready to countenance threats to the environment than to question general consumption patterns and citizen behaviour, and any decrease of consumer expenditure or GNP is greatly feared. In Beck's view, the massive profit and property interests that advance industrialisation and the growth economy have been in contradiction with the moral norm of reducing material consumption, and have blocked any political regulation of unsustainable lifestyles or general decrease in material affluence. Ultimately, what governments seem to fear most are not the ecological risks and hazards of mass consumption, but the possible side effects of a morally inspired limitation of consumption levels: a decrease in consumer demand, the effects of market collapses and an overall devaluation of capital (Beck, 2000).

Looking for an alternative 'third way': Towards a sustainable hedonism

We have seen that the hedonistic conception of the 'art of living' is ecologically, socially and culturally destructive. Anyone who has studied the *IPCC* reports published in 2001 and 2007 or read the many recent articles in prominent journals such as *Nature* and *Science* must come to the conclusion that current levels of production and consumption are simply unsustainable. As I have argued elsewhere, ultimately we shall be forced to reduce material consumption on a large scale, which in Western liberal democracies will imply a break with prevailing views on 'the good life' and the 'art of living' (de Geus, 2003).

Yet a strategy based on an essentially moralistic approach to the 'art of living' is also open to criticism. In Western liberal democracies citizens are accustomed to be free, in the sense of enjoying consumer choice, and directing their own lives without too many obligations and prohibitions. Overall, citizens prefer to live by their own views on 'the good life' and individual happiness. We thus have to ask in what the founding principles of a sustainable hedonism and an ecologically sensitive 'art of living' might consist. Is there a third way alternative, and if so, what are its main constituents and commitments?

First, we should not aim to attain to some perfect or ideal form of the 'art of living', or state of 'the good life' and human happiness. The history of political philosophy has taught us that it is inadvisable to strive for ultimate perfection of that kind. According to Karl Popper, our knowledge and experience are too limited to allow the state to implement any very radical changes in the organisation of society or the lifestyles of individuals. For him the so-called 'utopian' or 'idealistic' approach is subject to huge uncertainties and risks: undesired results will be the consequence, as will unexpected developments and discrepancies between theory and practice (Popper, 1974, p. 167ff). It is far wiser to opt for a pragmatic strategy, one which bases its interpretation of an alternative and viable (because sustainable) 'art of living' on real-life situations, rather than utopian wishfulness. Such an approach favours flexibility and intelligent adaptive behaviour as the preferred means by which citizens would learn to live well and to adopt more socially responsible habits.

Secondly, anyone wanting to shift views on the 'art of living' will have to acknowledge that modern society cannot thrive without a considerable divergence of lifestyles and that there is a general right to human development and fulfilment. In the past there have admittedly been strands of green political thinking that tended to prescribe very determinate ways of life and proposed a radical simplification of lifestyles. However, nowadays this overly narrow-minded approach has been replaced by more realistic and freedom-oriented strategies which take into account the blessings and comforts of modern society.

It is a fact of life that in Western liberal democracies freedom and diversity of lifestyle, variety in general, and the existence of numerous possible forms of self-expression are considered among the highest social values. We have to accept therefore that ecologically sound lifestyles will take many and various forms, and that sustainability and biodiversity can, and should, go hand in hand with lifestyle diversity. Fortunately, nowadays there is a widespread tendency in society to believe that since there are many forms of sustainable societies imaginable, there also exist many environmentally responsible lifestyles and 'arts of living' (Dobson, 2003, p. 21ff; Wissenburg, 1998, pp. 202–205).

Thirdly, Western liberal democracies not only have responsibilities towards those already alive, but also towards now highly vulnerable future generations. There is therefore a case for claiming that governments should these days be asking for some kind of 'sacrifices' from contemporary consumers, more precisely, that they should decrease their levels of material consumption and be prepared critically to rethink and reformulate their established visions of 'the good life'.

Fourthly, it has now become evident that changes in the interpretations and perceptions of 'the good life' and the 'art of living' are not only advisable for ecological, but also for social, cultural and psychological reasons. International empirical research into feelings of happiness has concluded that materialist lifestyles do not positively correlate to feelings of felicity and bliss. The increasing consumption and high levels of material affluence of the last 50 years in the Western world have not, by and large, made people happier (Kasser, 2002, pp. 22, 59, 72; *World Database of Happiness*). Critical social thinkers such as Erich Fromm, Ivan

Illich and more recently American sociologist Juliet Schor and psychologist Tim Kasser have pointed out that once people are above poverty levels of income, an increase in wealth has no positive effect in terms of happiness or well-being. In Kasser's argument, 'beyond the point of providing for food, shelter, and safety, increases in wealth do little to improve people's well-being and happiness' (Kasser, 2002, p. 47). On the contrary, materialism and the aspiration to high consumption usually leads to increased stress, rather than greater happiness.

Being versus having

What, then, are the constituents of an alternative approach? I am firmly of the opinion that contemporary patterns of consumption pose what is essentially a social and cultural problem relating to an 'outward-looking' and emulative concept of human well-being. Our Western liberal culture is too focussed on outward appearances, status and success. Following Erich Fromm, one can argue that in modern times we increasingly live in a society of 'having', instead of 'being'. Where 'having' provides the context of our existence, the emphasis is on the possession of material goods and things (Fromm, 1976); where 'being' does so, our feelings and aspirations are directed to other aspects of well-being more directly associated with our personal relations with others (care, love, friendship, self-realisation, environmental concern).

The continuous striving in Western culture for the satisfaction of material needs and the acquisition of goods has diminished the attention given to interactive and spiritual aspects of existence. Here, then, is the key to reducing levels of consumption while yet enriching lives and allowing people more pleasure and fulfilment. In a post-consumerist society the main emphasis will not be on the outward manifestations of status and success, but on cultivating the inward aspects of human well-being. One might cite here the pleasures of relaxation and balance, of attending more closely to our fellow creatures (both human and non-human), the enjoyment of meaningful work, of contributing to the community and of a general spiritual well-being: of maintaining a dignified, relaxed and elegant lifestyle, instead of being constantly drawn into the accumulation of possessions (Fromm, 1976).

In recent decades this fundamental insight has indeed been confirmed by extensive empirical sociological and psychological research. The wish to possess more and more material goods often places heavy burdens on modern citizens. Uncertainty and stress are significantly more common among materialistically oriented people, and so, too, are feelings of alienation and mental depression. Materialists have to work harder and longer hours to be able to purchase all their new goods, and also to maintain, repair and replace them at the end of the ceaseless cycle of work and spend. Because of this, they have less energy and time to spend on the most elementary forms of well-being: living, caring, loving, learning, and the enjoyment of sport, culture and other recreations (Kasser, 2002, pp. 97–116).

Balance of life and the importance of human fulfilment

At the beginning of this chapter I cited Hannah Arendt (1906–1975) as a prominent modern thinker on 'the good life' and the 'art of living'. In *The Human Condition*, she argues that there are three fundamental human activities: labour, work and action. Labour is the activity which is linked to the biological processes of the human body and the satisfaction of direct physical needs, like the provision of food and drink. Work is the activity that creates an artificial world of enduring objects, in order to provide a stable living environment for humans. Action corresponds to the fundamental human condition of plurality and is directed at the creation of a public sphere in which people can enjoy political freedom (Arendt, 1985, pp. 7, 50ff).

Arendt contends that while action was of the highest significance in the Greek polis, in ranking followed by work and finally labour, in modern times this hierarchy has been completely reversed. Political changes resulting in the systematic loss of the so-called 'spaces' of political action and the erosion of the distinction between the private and the public realms have contributed to this reversal. Nowadays, the value of labour and the production of consumption goods are consistently overrated, whereas the value of the more meaningful activities of acting in the public realm and creating lasting products is scarcely recognised (Arendt, 1985, pp. 294–305). This means that the goal of human activities (*vita activa*) is no longer to be found in sustaining a public sphere for political action or in creative work, but in routine labour that advances economic growth, abundance and mass consumption. Nowadays, expanding wealth and promoting the greatest consumer happiness of the greatest number are the basic aims of social and political life (Arendt, 1985, p. 133). With this conclusion, Arendt offers the key to a new understanding of 'the good life' and a green 'art of living'. For Arendt, the central issue in the debate is not primarily about the need for an immediate reduction in levels of consumption and an end to the waste economy.

Instead, she calls for reflection on 'the human condition' by maintaining a delicate balance between the activities of life. She aspires to a balanced configuration of the three basic components of the *vita activa*. In the words of Kerry H. Whiteside's instructive article on the subject:

> An active life combines labor, work and action in a unique configuration. Labor both produces and respects a worked-upon world of more permanent creations; this world in turn, derives its significance from public deliberation and memorialisation. Deliberation needs labor's energies in order to maintain its sense of life; it needs work's accomplishments in order to appear. Maintaining the proper ordering of human existence is a matter of correctly integrating the three activities. Life becomes distorted if it is given over exclusively to laboring and consumption or to work and instrumentalisation, or to deliberation and public display. Thus, adhering to this irreducible pluralism of activities leads us away from a life dominated by labor and consumption that – as many ecological political thinkers have concluded – is incompatible with the welfare of all living things.
>
> (Whiteside, 1994, p. 354)

All in all, this indicates a fundamental shift in thinking about 'the good life' and a green 'art of living'. It suggests that human fulfilment and the living of a rewarding and well-spent life can best be approached from the perspective of a proper and balanced ordering of action, work and labour, these being the basic components that constitute a happy, satisfying and virtuous human life.

Ecological hedonism

To be sure, this approach to a sustainable lifestyle and an ecological 'art of living' does not exclude every form of hedonism. The pleasures associated with consumerist acquisition are ignored here, but not those which cause no serious environmental pollution or degradation.

As the Dutch social philosopher Ton Lemaire has told us in his inspiring *Met Open Zinnen; Natuur, Landschap, Aarde* ('With Open Senses; Nature, Landscape, the Earth'), there is a more authentic hedonism distinct from that associated with the superficial seductions of the world of goods and consumer society. This is what he calls a 'spiritual naturalism' which rejoices in the pleasures of nature. For example, our senses enable us to have intense feeling of happiness when looking at a beautiful tree, a bird circling in the sky, or a butterfly in our garden. As Lemaire explains:

> We can smell the odour of a flower, feel the bark of a tree or the smoothness of a rock; the lightly sour taste of an apple, the sweetness of honey on the tongue. We can enjoy the singing of a blackbird on a spring morning, the rustling of the wind or the rippling of water; our eyes can rejoice in the many nuances of the green of trees and bushes, the vivid yellow of sunflowers or the simple dandelion, the brown tones of the earth. While walking we appreciate the world as landscape, admire panoramas and find bliss in looking at sublime views.
>
> (Lemaire, 2002, p. 272, my translation)

Such experiences are the gift of nature, incurring no financial expense. Hedonism of this kind, as Lemaire acknowledges, is enjoyment of sensual pleasure, but in ways that respect the natural environment. This kind of hedonism is in my view best called 'sustainable hedonism'. It is essentially spiritually oriented, leaves nature unharmed, does not disturb ecological balance and is not obsessed with the relentless acquisition of novelty and luxury.

Limits to our ecological footprint

Above, I argued that a great diversity of lifestyles is a fact of modern life. However, what is also needed is a fundamental re-examination of the proper 'ecological limits' of the variety of lifestyles and 'arts of living' which are currently enjoyed. In the end the exact shape that people give to their lives and their specific ideas of the 'art of living' are less important than their commitment – regardless of the specific lifestyle they choose – to living within the borders set by the carrying capacity of the Earth. By whatever means, individual citizens should now aim to keep their so-called 'ecological footprint' within sustainable limits.

As Mathis Wackernagel and William Rees have noted, the ecological footprint is a measure of the land area needed to sustain the levels of resource consumption and waste discharge by the population of a country (Wackernagel and Rees, 1995, p. 51ff). Recent footprint analyses show that the available ecological space on Earth is approximately 1.8 hectares per head, whereas the ecological footprint of people in rich countries like the United States (9.5), Australia (7.7) and Canada (6.4) has expanded massively (Jones and Jacobs, 2006, p. 216). If every person in the world had an American, Australian or Canadian standard of life, we would ultimately need three or even four Earths (Jones and Jacobs, 2006, p. 214).

It is relatively easy to estimate one's average ecological footprint by filling in a simple questionnaire on the Internet (see, for example, http://www.footprint.org). In fact, it only takes a few minutes to find out for yourself which forms of consumption add most to your impact on the Earth (For example, household gas consumption, car use, flying, electricity use, high meat consumption). But government could also improve public information through the use of 'eco-labelling'. This would indicate the ecological footprints of certain goods relative to the average 'footprint' available to humans by giving details about the land and energy used, and carbon dioxide emitted, in their production and consumption.

The real question concerns the extent to which, individually and collectively, we come to acknowledge the impact of our behaviour on the Earth and start to adjust our everyday material consumption in the light of global ecological constraints. This would require us to understand the approximate amount of environmental space that we use and to be prepared to reduce our overall claims on land, energy and other natural resources. This is logically implied by the call from Wackernagel and Rees for a so-called 'fair Earthshare' of 1.8 hectares for every inhabitant of the planet. They define this as 'the amount of land each person would get if all the ecologically productive land on Earth were divided among the present world population' (p. 53). In sum, provided that people do not irresponsibly exceed their fair share of the Earth, an ecologically responsible lifestyle need not be one-dimensional or bleak, but can offer its own rewards and excitements.

The cultivation of ecological virtues

In line with the analysis so far, a central prerequisite of a sustainable hedonism and a green 'art of living' is the cultivation of 'ecological virtue', as suggested by my Irish colleague, green theorist John Barry. He convincingly rejects the claim that an ecologically balanced society can only be achieved through a complete disavowal of materialistic lifestyles, or extensive self-denial or asceticism. There is no good reason to condemn all forms of consumption of goods and services in our world. Barry does, however, stress the need to explore the cultivation and development of green virtues and a 'greened' moral character:

> To use ecological terminology, virtues may be thought of as character traits, modes of being which help to find the best 'adaptive fit'

> between the individual and her interests and the environment (both social and natural) she inhabits. The importance of green virtues for the green position resides in the necessity of self-restraint, prudence and foresight so that long-term (i.e. sustainable) well-being is not sacrificed or undermined by desires to satisfy immediate self-interest.
> (Barry, 1999, p. 35)

In fact, ecological virtues may correct human frailties and in particular prevent inaction or 'weakness of will' (in Greek: *akrasia*), in other words, overcome the gap between knowledge/belief and action. Many of us know what we should do (consume and travel less by air, be less selfish and so on) but lack the will (energy, courage) to do so. In the words of Barry, the general idea behind this approach is that 'the cultivation of the ecological virtues, the creation of ecological character and dispositions, help create and maintain a proper balance within social-environmental relations. The emphasis on character stresses the importance of cultivating dispositions and modes of action which will discourage acting from wantonness or ignorance' (p. 67).

Barry rightly argues for the development of a sophisticated theory of ecological virtues and moral character. Such a project raises key issues concerning the general attitudes of humans towards nature, the importance of material consumption in human life, and how we can live 'the good life' in responsible ways without endangering long-term sustainability. Rather than focusing only on broad and impersonal social concepts like a 'stable state society' or a 'green market economy', the notion of ecological virtues (and of course vices) concentrates on individual attitudes and actions, on our fundamental moral ideals and personal views on living 'the good life'. In principle, virtues like prudence, respect, care, moderation, self-control, tolerance, balance of activities and awareness of limits to growth and material welfare would seem to be of vital importance here.

Ecological citizenship and ecological consumerism

Finally, one of the most radical changes in current conceptions of 'the good life' and a green 'art of living' in times of worldwide environmental crisis is to be found in the area of ecological citizenship and ecological consumerism. In line with green political theorists such as Andrew Dobson, I would like to describe such citizenship as one in which the obligations and responsibilities of humans 'extend through time as well as space, towards generations yet to be born' (Dobson,

2003, p. 106). Ecological citizens have learned to take into account their moral and environmental obligations and responsibilities towards other humans on earth and future generations.

Ecological consumerism can be described as a mode of behaviour in which the selection and purchasing of products and services is not based on considerations of utility maximisation, individual profit and enjoyment, but on considerations and feelings of social responsibility and the burden placed on our environment. Politics would then be licensed to interfere increasingly in the buying behaviour and consumption options of citizens as consumers, since ultimately an overall reduction of levels of material consumption is an important strategy in the abatement of environmental pollution and the realisation of ecological sustainability in the world.

Conclusion

The problem outlined at the beginning of this chapter is how the world's citizens can all live comfortably and well, within the limited ecological means of the earth. In order to find an answer to this question, I first analysed the two dominant approaches to 'the good life' and a green 'art of living': the moralistic and the hedonistic. The hedonistic line of thinking, which is based on the satisfaction of material needs and the quest for luxury and pleasure, was shown as leading to worldwide environmental problems such as climate change, scarcity of resources and a decline of species. It was also shown that modern hedonism has lent itself to an ideology which promotes possessive values and materialistic attitudes, and means that the gap between desire and fulfillment can no longer be bridged.

There are a number of dangers and disadvantages attached to the moralistic approach, including the contemporary preference of individuals to pursue their own conceptions of 'the good life' and their distaste for any ethically driven government interference in the determination of their lifestyles. An alternative approach, I have suggested, should be based on pragmatism, respect for the diversity of lifestyles, feelings of responsibility towards the present and future generations, and the basic empirical insight that abundance does not make people happier.

A 'sustainable hedonism' requires a shift from 'having' to 'being', and a balanced configuration of the action, work and labour components of the *vita active*. This is compatible with forms of hedonism that differ from those of mass consumerism. In promoting it, we need to relate our material consumption to our ecological footprints, and also to cultivate ecological virtues and moral character. Ecological citizens and consumers will have to learn to live in responsible ways and to take notice of their fellow humans and future generations.

An ecological 'art of living' implies a permanent learning process that seeks to integrate individual freedom and diversity of lifestyles with social-ecological and moral responsibilities towards our fellow humans and descendants. This can lead to a sustainable hedonism: an alternative hedonism which is both enjoyable and rewarding without exceeding our fair shares of the earth's resources.

References

Arendt, Hannah (1985) *The Human Condition* Chicago: The University of Chicago Press.

Aristotle (1975) *Ethics* Harmondsworth: Penguin.

Barry, John (1999) *Rethinking Green Politics* London: Sage Publications.

Beck, Ulrich (2000) *Risk Society: Towards a New Modernity* London: Sage Publications.

Campbell, Colin (1989) *The Romantic Ethic and the Spirit of Modern Consumerism* Oxford: Basil Blackwell.

Dobson, Andrew (2003) *Citizenship and the Environment* Oxford: Oxford University Press.

Dohmen, Joep (2002) *Over Levenskunst: De Grote Filosofen Over Het Goede Leven* Amsterdam: Ambo.

Fromm, Erich (1976) *To Have, or to Be?* New York: Harper and Row.

Geus, Marius de (1999) *Ecological Utopias: Envisioning the Sustainable Society* Utrecht: International Books.

Geus, Marius de (2003) *The End of Over-consumption: Towards a Lifestyle of Moderation and Self-restraint* Utrecht: International Books.

Hobbes, Thomas (1974) *Leviathan* Harmondsworth: Penguin.

Jones, Peter Tom, and Jacobs, Roger (2006) *Terra Incognita* Gent: Academia Press.

Kasser, Tim (2002) *The High Price of Materialism* Cambridge, Massachusetts: Bradford Books.

Lemaire, Ton (2002) *Met Open Zinnen: Natuur, Landschap, Aarde* Amsterdam: Ambo.

Locke, John (1965) *Two Treatises of Government* New York: Cambridge University Press.

Nehamas, Alexander (2000) *The Art of Living: Socratic Reflections from Plato to Foucault* Columbia and Princeton: University Press of California.

Nussbaum, Martha, and Sen, Amartya (eds) (1993) *The Quality of Life* Oxford: Oxford University Press.

Pascal, Blaise (1995) *Pensées* Harmondsworth: Penguin.

Popper, Karl (1974) *The Open Society and its Enemies*, Vol. 1, London: Routledge and Kegan Paul.

Rousseau, Jean-Jacques (ed. V. Gournevitch) (1997) *The 'Discourses' and Other Early Political Writings* Cambridge: Cambridge University Press.

Veenhoven, Ruut (2002) 'Leuk Levenskunst', in Dohmen, Joep (ed.) *Over Levenskunst: De Grote Filosofen Over Het Goede Leven* Amsterdam: Ambo.
Wackernagel, Mathis, and Rees, William (1995) *Our Ecological Footprint: Reducing Human Impact on the Earth* Gabriola Island: New Publishers.
Whiteside, Kerry H. (1994) 'Hannah Arendt and Ecological Politics' *Environmental Ethics* 16(4), pp. 339–358.
Wissenburg, Marcel (1998) *Green Liberalism the Free and Green Society* London: UCL Press.
World Database of Happiness, www.eur.nl/fsw/research/happiness (accessed 24 August 2007).

7
Green Pleasures

Richard Kerridge

> So swimming is a rite of passage, a crossing of boundaries: the line of the shore, the bank of the river, the edge of the pool, the surface itself. When you enter the water, something like metamorphosis happens. Leaving behind the land, you go through the looking-glass surface and enter a new world, in which survival, not ambition or desire, is the dominant aim. [...] You see and experience things when you're swimming in a way that is completely different from any other. You are *in* nature, part and parcel of it, in a far more complete and intense way than on dry land, and your sense of the present is overwhelming.
>
> (Deakin, 2000, pp. 3–4)

This is from *Waterlog*, Roger Deakin's narrative of how he took a 'swimming journey' across Britain, his route including much 'wild swimming' in water where swimming was not allowed, or not expected. I offer it as a first example to help define environmentalist hedonism. Deakin, who died in 2006, had played a founding role in Friends of the Earth and Common Ground, as well as being a nature writer and environmentalist film-maker.

Enjoying nature

'The ecologically impaired', suggests Andrew Ross, 'need to be persuaded that ecology can be sexy, and not self-denying' (Ross, 1994, p. 15). He speaks of 'the hedonism that environmentalist politics so desperately needs for it to be populist and libertarian' (p. 17). Kate Soper has observed that 'it is abundantly clear that... anxieties in themselves have not proved sufficient to prompt any radical transformation of consumer habits'. She agrees with Ross about the need for 'an alternative hedonist vision' involving 'very different conceptions of consumption and human welfare from those promoted under capitalism' (Soper, 1995, p. 271). Is Deakin's 'wild swimming' sexy? Does it compare, for example, with Colin Firth, as Mr D'Arcy, plunging Romantically into the lake in *Pride and Prejudice* on television: a recent icon of Romantic sexiness?

Deakin's strange journey is an eccentrically adventurous example. What are the more obvious 'green' pleasures? 'Slow food' comes to mind, as do hiking and wildlife-watching (in the wild and on television). The love of nature is still a strong presence in popular culture. Historically it is by no means an exclusively middle-class enthusiasm (see Matless, 1998, pp. 70–79, for example, on the history of socialist rambling clubs and on the Kinder Scout trespass). TV wildlife is one of the most reliably popular genres. But eco-tourism has to be 'costed' in environmental terms. Safari holidays may gratify a love of nature, but how much carbon dioxide did the journey release, and does the tourist industry have a benign or damaging effect on local ecosystems? Eco-hedonists need to be willing to ask such questions. If loving nature is to become central to popular culture in a new way – as environmental crisis demands – it will be integrated with new and restrictive conditions. 'Green' hedonism must remind us that there are other freedoms, other forms of expansiveness, beside carbon-guzzling consumerism.

A 'green' pleasure is one that follows the logic of environmentalism – by using less carbon, deepening one's love of things already at hand, appreciating cycles of growth and renewal in the local and global ecosystems, understanding and taking delight in interdependency. This logic might point, for example, to taking joy in family and friendship, rather than working long hours (if you get the choice). It points to holistic care for the body rather than binges, addictions and quick fixes for symptoms. With this goes affection for one's naturally formed and naturally ageing body, and resistance to the commercialised 'beauty myth'. And there should be 'green' pleasure in craft rather than mass production; making things rather than buying them; knowing and repairing them, treating them even with a kind of loyalty, rather than discarding them as soon as they are old or impaired.

These are the sorts of pleasure that are derided as 'smug', 'worthy' and 'pious', smug self-approval being the wrong sort of contentedness at living within limits. To counter these sneers, environmentalists need to demonstrate a capacity for humour, irony and restless self-critique, as well as serenity; a capacity to find the infinite as well as to be content with limitations. Countering the wastefulness and insatiability of consumerism might mean reasserting the pleasure of feeling 'I have enough', but 'enough' must have infinite depths: always more to explore.

Deakin's 'wild swimming' is a defiance of 'a world where more and more places and things are signposted, labeled and officially "interpreted" ' (Deakin, 2000, p. 4). From the environmentalist point of view, this raises questions. Ecological management and energy-saving require a lot of measuring, forecasting and rule-making, leaving little room for wild heedlessness. Deakin knows this, and understands the paradoxical quality of present-day environmentalist enthusiasm for wild nature. Wildness: prized for its freedom, its mystery, its secret, intimate places, its capacity to surprise; yet protected, understood, watched-over, managed and, in its meanings, culturally relative. Anyone who combines Romanticism with scientific interests encounters this paradox. Deakin's recognition of it is shown by the importance of thresholds and liminality to his hedonism. The pleasure comes from a love of wildness: adventure, unpredictability, the unknown, the sudden encounter, the absence of rules and restraint. That is wild nature. But now nature must be *knowable* as never before: this is a secondary cost of our environmental crisis, as Bill McKibben saw back in 1990:

> We have deprived nature of its independence, and that is fatal to its meaning. Nature's independence *is* its meaning; without it there is nothing but us.
>
> (McKibben, 1990, p. 54)

How can the two sides of this paradox be reconciled? Deakin shows one way. He lowers himself into patches of water where no one swims. They are familiar parts of the landscape, but it is as if he has discovered an unnoticed fold: it does not interrupt the smooth surface of the known environment, but wildlife lurks inside. His journey is a good example of a certain kind of defiant enjoyment of nature that exults in the fact that even such a country as Britain, with its densely populated and domesticated environment, still has, all over the place, small strips and patches of natural wildness. The book is in a nature writing tradition that the poet Jeremy Hooker has called 'ditch vision'; his name for the joyful experience of nature that can flourish in a mainly urban and suburban country. The British seeker of Romantic solitude in nature, especially in England, will have difficulty in finding any place of wild seclusion that is not liable to disturbance by other walkers or nearby traffic. There are wonderful moors and mountains, and some surprisingly wild and lonely places, but only the Scottish Highlands offer anything that begins to resemble North American 'wilderness' (whatever problems that term, in its purist sense, raises).

Liminal stages through which the traveller passes, going in deeper, forgetting the normal and reorienting to the wild, are not available, unless they can be achieved by more purely imaginative means. Nature and seclusion have to be found in the midst of society. 'Ditch vision' names the imaginative habit of playing with scale in order to discover wildness and infinity in small spaces; the genre of daydreaming that sees in an overgrown railway bank the principle and possibility of wildness. The term also conjures the absorbed gaze of a child to whom the local world has not yet come to seem small and familiar. There is a perception now that traffic congestion, over-protective parents and addiction to television and computers are cutting children off from direct experience of wild nature, even of the 'ditch vision' type. I am afraid this lack now goes by the name of Nature-Deficit Disorder. For an account of it, see Richard Louv's *Last Child in the Woods* (2005).

Childhood, in the Romantic imagination, was to be contemplated with joy because children, trailing clouds of glory, retained for a few precious years their sense of living in an infinite world. In adulthood, in general, and especially in modern, urban life, this sense of the infinite fades. Wild nature was where it might be discovered again. The finding of infinity in a small and ordinary place is, in the Wordsworthian tradition, an antidote to the influence of reductive and Cartesian science; the science that treated living organisms as knowable mechanisms.

Essential to the appeal of swimming for Deakin is the crossing of a precise threshold, the plunge into a different element, accentuated when the swimming is 'wild' and the water is not clear and chlorinated but dark and full of newts. Just as important is the emergence, the crossing back into conventional perspective and accountability. Some wilderness writing is fascinated by the possibility of venturing beyond the last point of return (Jean Hegland's novel *Into the Forest* (1996) is a recent example); passing over into a state of wildness or unselfconsciousness from which messages cannot reach us. Such proximity to a point of no return is a version of the Romantic sublime. But environmentalists need to move freely and frequently between the enraptured contemplation of wild, infinite nature, and the pragmatic measuring and contextualising that the crisis makes necessary.

Another contemporary nature writer who describes the liminality of swimming is Jay Griffiths, in *Wild*. Here she writes about tropical ocean diving:

> For the human mind, the crossing from air down into water has powerful effects. Unable to smell, losing accustomed gravity and losing our air hearing, it is surprisingly difficult to carry memory across the border...The air world is hard to recall on the instant of merging in the saltwater world we knew before...Underwater, mind belongs. Our earliest perceptions as both creatures and individual fetus are of gentle thrilling waters.
>
> (Griffiths, 2006, pp. 164–165)

Yet the immersion and – as she imagines it – regression has a strict time limit placed on it by her oxygen supply. The line of demarcation between this subliminal world and the reality she normally inhabits is absolutely clear: no danger of not being sure which she is in. Her experience is very precisely bordered.

These writers savour the border territory; the experience of transition, and the first shock of immersion, when the senses are alienated and surprised. But what of environmentalist hedonism that goes further in than the border territory, and is prepared to lose touch with the external perspectives?

Ecophenomenological pleasure: David Abram

Is there a way of experiencing the world that would make environmentalism a natural consequence of ordinary life rather than a difficult programme of self-denial?

Can we imagine a reorientation of our bodily senses that would make our bodies *want* to take the actions necessary to avert catastrophic climate change? Could we come to act not only because we have been persuaded of the need, but because our physical impulses and appetites lead us that way? Acting to avert climate change would then not be a matter of restraining our impulses but of releasing them. This may seem improbable, but perhaps less so if we believe our bodies may already be attempting, constantly, to connect with the natural world, while we barely know it, being so alienated from them. Theories of our modern alienation from our natural selves feature in most kinds of environmental thinking, but it is in the phenomenological tradition that environmentalists have found this enticing possibility of a reorientation of the senses. They find this especially in the work of Martin Heidegger and, more recently, Maurice Merleau-Ponty.

Here is David Abram, the best-known popular advocate of the ecophenomenological approach:

> Caught up in a mass of abstractions, our attention hypnotized by a host of human-made technologies that only reflect us back to ourselves, it is all too easy for us to forget our carnal inherence in a more-than-human matrix of sensations and sensibilities. Our bodies have formed themselves in delicate reciprocity with the manifold textures, sounds and shapes of an animate earth – our eyes have evolved in subtle interaction with *other* eyes, as our ears are attuned by their very structure to the howling of wolves and the honking of geese. To shut ourselves off from these other voices, to continue by our lifestyles to condemn these other sensibilities to the oblivion of extinction, is to rob our own senses of their integrity, and to rob our own minds of their coherence. We are human only in contact, and conviviality, with what is not human.
>
> (Abram, 1997, p. 22)

'Delicate reciprocity with the manifold textures, sounds and shapes of an animate earth': this sounds pleasurable and proposes the reintegration of sensuous pleasure into other forms of practical engagement with life – so that Abram's idea is not, strictly, a hedonism; it does not separate the delectation of the senses from all their other purposes.

The Spell of the Sensuous, published in 1996, is the most accessible statement yet of the ecophenomenological approach. Phenomenological philosophy has often, paradoxically enough, been abstract and difficult to read. It seeks to awaken our senses and banish abstraction by persuading us with a lot of highly complex abstract argument. But Abram is not difficult to read. He places his philosophical argument in the context of a personal story, a secular version of the 'born again' narrative of revelation and conversion, with an appeal beyond professional academic circles. The book can be found in the self-help and alternative healing sections of bookshops as well as the philosophy and environment sections.

His principal philosophical source is Merleau-Ponty, who identifies the experiencing subject with the physical body rather than with the self-conscious mind considered apart from the body and felt to be separate from it, as in the Cartesian tradition. 'Embodied' perception, as Merleau-Ponty proposes it, involves abandoning the notion of things as mere *objects* of perception. Instead, he gives an account of perception as a dynamic relationship, in which the perceiving body is constituted in the act of perception. Abram summarises:

> Perception, in Merleau-Ponty's work, is precisely this reciprocity, the ongoing interchange between my body and the entities that surround it. It is a sort of silent conversation that I carry on with things, a continuous dialogue that unfolds far below my verbal awareness – and often, even, *independent* of my verbal awareness, as when my hand readily navigates the space between these scribed pages and the coffee cup across the table without my having to think about it, or when my legs, hiking, continually attune and adjust themselves to the varying steepness of the mountain slopes behind this house without my verbal consciousness needing to direct those adjustments. Whenever I quiet the persistent chatter of words within my head, I find this silent or wordless dance always already going on – this improvised duet between my animal body and the fluid, breathing landscape that it inhabits.
>
> (p. 52f)

Deakin and Griffiths describe literal acts of immersion that shock the senses into life – an alienation of the senses comparable in literary terms to the device of 'making strange' that describes familiar things as if we did not know what they were, and in phenomenological terms to Husserl's 'époché', or suspension of presuppositions before an encounter. Abram takes us as 'always already' immersed but mainly unaware of our sensuous immersion in the world. Modern life has dulled our appreciation of it. By silencing our pre-emptive verbal thoughts – his version of the époché – we can begin to rediscover our bodily life in the world, and find in the most ordinary actions an enhanced reality.

Living with villagers in Indonesia and Nepal, Abram found he was becoming more and more responsive, in all his senses, to the natural world. He found himself noticing more detail in the natural world around him, and hearing it, smelling it and feeling its touch. Inspired by these experiences, *The Spell of the Sensuous* exhorts us all to set out to find those forms of awareness, cultivate them and bring them to life, not only on visits to non-industrial societies but at home.

Abram's experiences were not practical components of daily life but isolated, epiphanic moments of encounter. This is what being modern does to such experiences. Exiled from ordinary practical life, they are discovered in the climactic moments of adventures away from home or, on a smaller scale, in moments of dreamy idleness. To describe this repositioning is to make a good start at summarising Romanticism.

Ecophenomenologists are certainly not advocating an abandonment of modern technologies in favour of a new hunter-gatherer subsistence. But if all the inspiring examples are drawn either from leisure or from indigenous cultures, and there is no suggestion that modern cultures should return to some form of indigenity, then ecophenomenology is revealed as a renewed form of Romantic wistfulness; its great pleasures, the awestruck gaze and the intimation. Perhaps Abram too does not get beyond the border zone.

What are the problems with this radical vision of 'green' life as richly pleasurable rather than austere? One is that Abram's writing – and writing of this kind generally – is signally lacking in one kind of pleasure: humour and self-irony. This is not surprising. Irony involves stepping back from oneself – disclaiming oneself, even – in order to watch and be amused. Abram seeks to go so deep into embodied life that such a division would be impossible. But the absence of humour is troubling, and makes his work vulnerable to the 'smug' and 'pious' line of attack. A greater sense of relative positioning might help: and of the desirability of moving into embodied perception and then sometimes retreating back to ironic distance. This would be to have it both ways: to combine being neo-indigenous with being modern. And as long as Abram is still reporting back to us, we know he has not reached the condition of undivided embodiment yet. To go in so deep would also be incompatible with retaining the large scientific perspective, the measuring, the long-range forecasting that environmentalism requires – and of which, in many ways, it consists.

So Abram's is a millenarian vision, as befits the religious genres he adapts to his purpose. As with all conversion narratives, nothing significant can happen until everything happens, in one revelation. His radical revolutionary narrative is opposed to gradualist, incremental change. But there are other versions of this approach.

Ecophenomenology without revelation: Tim Ingold and Wendy Wheeler

In *The Perception of the Environment* (2000), the social anthropologist Tim Ingold's project is to bring phenomenological ideas, notably those of Heidegger and Merleau-Ponty, to bear upon his discipline. Although his aims are not specifically environmentalist, he frequently uses the term 'ecological' to characterise his quest for a way of experiencing and describing landscape, and human life in landscape, that is sensuously alive to the manner in which the landscape and the life are continuously producing each other. He rejects what he sees as 'a systematic bias in Western thought': the tendency 'to privilege form over process' (Ingold, 2000, p. 198). A painting, for example, will almost always be discussed as a final product rather than in terms of the work of painting. Ingold contrasts this with the practice of some non-Western cultures in which it is the activity of painting that constitutes the contemplative experience. The finished works are not kept, nor given any lingering attention.

Ingold uses the term 'taskscape' rather than 'landscape' in order to remind us that when our surroundings seem to be laid out for our gaze, our perception is a function of the activity we are engaged in, whether work or leisure. Perception is conditioned by the specialisation that a particular task involves – what the trained eye of the farmer or police officer picks out, as compared to that of the birdwatcher or historian – and by the duration of the task, its rhythms and the intervals for vision that it affords. 'Tasks', Ingold says, in Heideggerean vein, 'are the constitutive acts of dwelling'. And each task is not isolated, but 'takes its meaning from its position within an ensemble of tasks, performed in series or in parallel, and usually by many people working together' (p. 195). Drawing on the analogy with an orchestra, he argues that different 'tasks' within a system must be understood as always engaged with each other. Their being is to answer and modify each other, and they determine each other's rhythm and duration:

> Human beings do not, in their movements, inscribe their life histories upon the surface of nature as do writers upon the page; rather, these histories are woven, along with the life-cycles of plants and animals, into the texture of the surface itself. Thus the forms of the landscape arise alongside those of the taskscape, within the same current of activity. If we recognise a man's gait in the pattern of his footprints, it is not because the gait preceded the footprints and was 'inscribed' in them, but because the gait and the prints arose within the movement of the man's walking.
>
> (p. 198f)

The skill involved in that kind of recognition is clearly anachronistic, but Ingold cites it to show how the action of our lives continually produces our selves as well as the landscape we live in. Standing back from my life to contemplate it is something I sometimes think I do, just as I sometimes position myself where I can gaze out across a landscape, but the detachment I achieve is slight at best and may be illusory. I have not really stepped off my moving life, any more than when I pause at a scenic outlook I have disengaged myself from the economic, cultural and ecological processes of the landscape in which I am walking. As Ingold says, having just quoted Merleau-Ponty, the 'notion that we can stand aside and observe the passage of time is founded upon an illusion of disembodiment' (p. 196). Ingold wants to dispel that illusion, by insisting on modes of perception that are functions of tasks and not really separable from those tasks.

It is telling, though, that his main examples of tasks are non-industrial and, in the Western context, pre-modern. Part of Ingold's anthropological purpose is to make sure he is not interpreting non-Western cultures merely in Western terms. Anthropology is a case in which trying to do what phenomenology recommends – to render explicit and then escape the presuppositions that mediate experience – is vital to the integrity of an academic discipline; to its efforts to shake off its colonial past. Ingold would have his Western, modern readership able to grasp and respect the modes of perception of non-industrial cultures. But his further and more implicit goal is a phenomenologically inspired transformation of the way his Western readers perceive their own environments. He is suggesting that Westerners would appreciate their own world better by rediscovering old ways of perceiving and making new ways.

Environmentalists are keen to combine the protection of natural ecosystems with the support for the indigenous communities in those ecosystems. Supporting these communities means helping them defend and assert their rights to land. It also means supporting their demands for modern social services and economic opportunities, and this produces the dilemmas that are part of the daily work of the environmental justice movement.

Of course, there are powerful reasons to be wary of celebrating indigenous cultures for their unalienated relation to nature. In the European colonial tradition, such celebration has often been an offshoot of the ideology that legitimated colonial conquest by identifying the victims with nature. Ingold and Abram want to avoid giving that kind of colonial Romanticism its latest form. They also want to avoid an associated error: nostalgia for the pre-modern that dishonestly undervalues and disowns the modern. Yet, avoiding these traps, they want to learn from indigenous cultures and use their example to bring to life something akin to them in Western culture.

The idea is that we would then discover, and rediscover, a different range of perceptions and relations from those that have been commodified by industrial capitalism. No doubt capitalism would soon begin to exploit them commercially, but in their initial discovery there might be a space of opportunity. The Marxist cultural geographer David Harvey has written of 'the desire, the longing and in some instances even the practice of searching for a space "outside" of hegemonic social relations and valuations' (Harvey, 1996, p. 230). For all kinds of anti-capitalist radicals this is an enduring quest. For Abram and Ingold, phenomenology seems to show that the other space or dimension of life is always with us already. The trick is becoming alive to it: breaking through.

To begin to realise this vision would be to move towards a culture in which the primary forms of pleasure were bodily and mental perceptions of the growth and interdependency of all creatures, human and non-human. There would be a love of complexity and intricate interrelationships – a presumption in favour of seeing and encouraging all forms of growth and potential, rather than competing with them or feeling threatened by them. In her recent book *The Whole Creature* (2006), the cultural critic Wendy Wheeler argues that complex, continuous symbiosis in nature provides the essential model that makes this transformation plausible. Like Abram, Wheeler sees the phenomenological tradition as an antidote to the ills and false assumptions that dominate modernity, but her interest is primarily in phenomenological creativity as a social, collective way of living. Her examples of phenomenological awareness do not come primarily from encounters with wild creatures outside the domain of modern human society. She attends to nature in a different way, by emphasising the connection she sees between ecological concerns, human well-being and 'biosemiotics'. This last term names those deep processes mentioned above: the evolutionary cellular symbiosis and complex bodily exchange with the world uncovered by scientists such as Lynn Margulis, Brian Goodwin and Candace Pert. Wheeler hopes that, as newer scientific understanding of complex symbiosis percolates into public culture, symbiosis will replace competition as the main evolutionary idea. A new phenomenological life of openness to otherness would then become a matter of habit and ordinary commonsense.

Like Abram, but in a more social context, Wheeler looks for an alert openness to bodily and emotional messages that we often do not hear because dominant Western notions of rationality give them little recognition or attention. She agrees with Abram that there is a destructive violence in the Cartesian view of objectivity as a matter of disengaging oneself from objects in order to see them dispassionately. She says this view 'has done untold damage in human affairs' (p. 50). This is not an anti-science position but an argument founded on a rejection of the excessive dominance that a certain kind of scientific thinking came to have in Western culture.

Biosemiotics, Wheeler argues, allows us to move towards an 'ecological' view in which creatures and things can only be understood in their continuous relations with each other. The ethical implications are made clear in the following passage:

> The 'objects' that we meet, and which afford us possible actions, are poorly understood when only understood as separate 'objects' along the lines of the old Cartesian divide. What the 'object' affords us, if our life is not to be a series of deaths, is, as biosemiotics suggests, the responsibility of responsiveness. But this responsiveness – which artists and writers have often expressed in terms of a feeling of responsibility towards their creativity, a kind of lived ethos – is a responsiveness to process which emerges in the ecological relation between self and other, or, more abstractly, between similarity and difference. This structure of attentive responsiveness to otherness – a fundamental creativity and openness to newness evident in life – is also at the heart of what, for humans, constitutes an ethical relation. (p. 135f)

This brings scientific ecology into potentially fruitful contact with Heideggerean ideas of Being and with Levinasian ethics. Wheeler's thesis is also very consonant with the 'biophilia hypothesis' of the biologist Edward O. Wilson, according to which human beings have an evolved affective affiliation to other living organisms that is manifest in a wide range of emotional reactions.

Romanticism, and romantic consumerism

When ecocriticism – literary and cultural criticism from an environmentalist perspective – emerged in the early 1990s, one of its main topics was Romanticism, the first great revulsion in modern Western culture against the separation of humankind from nature. Romanticism was the movement that first defined human experience in modernity as *alienated* experience, and the movement that proposed the love of wild nature as the way to heal that alienation, favouring holism and interconnection over abstraction and mechanism. William Wordsworth's poem 'The World is too much with us' contains possibly the most memorable anti-consumerist line in English literature: 'Getting and spending we lay waste our powers'.

However, in his influential *The Romantic Ethic and the Spirit of Modern Consumerism* (1987), the sociologist Colin Campbell has argued that the origins of modern consumerism itself can be found in the Romantic tradition. How can Romanticism be the source of both the problem and its solution?

Campbell seeks to explain the insatiability that is the distinguishing feature of modern consumerism: the never-ending desire for new purchases, and the rapid disillusionment with products once they have been acquired. For the maintenance of the present economy, this rapid turnover must continue. Any slowing means recession and crisis. For environmentalists, this species of desire is the tragic flaw. Campbell sees it as the product of 'modern autonomous imaginative hedonism', which he also calls 'self-illusory hedonism' and describes as 'characterised by a longing to experience in reality those pleasures created and enjoyed in imagination, a longing which results in the ceaseless consumption of novelty' (Campbell, 1987, p. 205). This modern hedonism is different from the traditional kind that consists in filling one's time with sensuous pleasure; and strikingly different from utilitarian materialism, in that the practical utility of objects is not the driving motive for desiring them. Indeed, they are routinely discarded while still perfectly useable.

In the pre-modern world, Campbell argues, emotions were not thought of as arising within people and motivating their behaviour, but as evoked by the external world. Fear was the proper response called up by certain things; joy or awe by others. This changed with modernity – with scientific rationalism, the decline of animistic beliefs, and the influence of Cartesian dualism, which separated the self-conscious human mind, possessed of free will, from the remainder of the world, seen as unconscious mechanism:

> This increasing separation of man from the constraining influence of external agencies, this disenchantment of the world, and the consequent introjection of the power of agency and emotion into the being of man, was closely linked to the growth of self-consciousness. Such a uniquely modern ability is itself the product of these processes, as, in becoming aware of the 'object-ness' of the world and the 'subject-ness' of himself, man becomes aware of his own awareness poised between the two.
>
> (p. 73)

Campbell's historical narrative tracks the consequences of this change as it was mediated through Protestant pietism, Eighteenth-century Sentimentalism, Gothic fiction and, most influentially, Romanticism. In the Romantic reaction against Newtonian and Cartesian rationalism, new value was attached to intuition, intimation and passionate sympathy. Romantics saw the natural world, and the human self, as organism rather than mechanism, and were thus able to give a new importance to ideas of continuous personal growth, stimulated by endless discovery and rediscovery – of the outside world and one's own inner self. Instead of the finitude of classification, dissection, itemisation of parts and categorisation of wholes, Romantics sought to appreciate the infinity of creatures and the world: the endlessness of new discovery and new becoming.

From being an attribute of external God, infinity now becomes a property of all life and experience: the infinity of chains of connection and circulation that link individuals, and of continuing processes of growth and decay, inheritance and bequeathing. William Wordsworth, in 'The Prelude', sees on a mountainside 'woods decaying, never to be decayed' (Wordsworth, 2004, p. 245). This is what, in the Romantics, seems proto-ecological. The interdependency and mutual shaping – what Darwinists would later call co-evolution – of all the world's creatures and components seems to merge them in a single organic substance. All kinds of flow and merging – of rivers, impressions on the mind, feelings, memories, sounds, airs and breezes – are instances and images of this substance. Coleridge, in 'The Eolian Harp', writes:

> O the one life within us and abroad,
> Which meets all motion and becomes its soul,
> A light in sound, a sound-like power in light
> Rhythm in all thought, and joyance everywhere –
> Methinks it should have been impossible
> Not to love all things in a world so filled.
>
> (p. 87)

This is the impulse that caused the Ancient Mariner, seeing the water snakes, to bless them unaware (Coleridge, 1997, pp. 155, 176), redeeming himself; a moment cited enthusiastically by several ecocritics and one that now seems to chime with Edward O. Wilson's 'biophilia hypothesis'.

Yet Romanticism was a preoccupation with unresolved alienation as well as with joyous reunion, and the intimations of immortality and continuity are usually epiphanies: brief glimpses, momentary revelations. A characteristic stance of Romantic subjectivity is that of the observer standing at a brink or threshold, gazing but holding back, not moving to cross. Subject and object are held separate. Romantic subjectivity is a state of heightened, often agitated, self-consciousness that contemplates unselfconsciousness, often with yearning, yet cannot really, or more than momentarily, lose itself back into unconscious being; into nature. Kate Soper has called this yearning contemplation 'the envy of immanence', and it produces complex mixtures of momentary joy, elegiac regret and ambivalence about self-consciousness.

When it is a matter of gazing at wild nature, the Romantic observer can take pleasure from a complicated ambivalence, enjoying the spacious leisure to admire and even envy Nature's unselfconsciousness, which is its lack of leisure. The supposed capacity of wild creatures to be purely identified with their actions, their bodily life, means that they are never leisured observers. Life does not afford them that space. They are always fully engaged. This unselfconsciousness is an essential part of Nature's meaning. Usually, there is no great pressure on us to question our position as separate observer of this; even though the freedom to admire in this way, open to whatever nature does rather than constraining it to a utilitarian purpose, is clearly a consequence of being at leisure ourselves, so that the question of how we came to be at leisure, at whose expense, should not be far behind.

In practice, the spectacle of wildness is in some degree contrived by manipulation. This does not exactly mean that it is not what it seems, because nature itself is innocent of this manipulation. Within the field of observation the creatures are behaving naturally. But, whether we are talking about a nature reserve, a television documentary or a poem, that field has been deliberately set aside, protected, even micromanaged, precisely to produce the spectacle of wildness – which is therefore contained within limits. The indignation sometimes aroused by the revelation that wildlife documentaries have 'faked' their effects is a product of this paradox. We always knew there was a combination of contrivance and wildness, but is it much more unequal than we thought, threatening to eliminate the wildness altogether? Similarly, our own encounter with wildness is probably to be kept within safe limits. We are not really going to trust our lives to our hunting and foraging skills and our ability to improvise warmth and shelter. Our visit to the wild is controlled and temporary, and our knowledge of this is an indispensable ingredient of the experience. It must not be so intrusive, though, as to take away the thrill, which in part consists of the fantasy that we are not just taking a day's hiking, or that the barriers keeping the zoo animals away from us are not there. 'Ditch vision' involves the same ability to suspend our knowledge of the limitations of space. 'Extreme' sports, on the other hand, are those that contrive within the space of leisure the real hazards of contingent nature.

All this contrivance and simulation may be the best way we have of satisfying emotional and physical – some would also say spiritual – faculties that industrial modernity has left unused. The best, that is, in the absence of modern forms of life capable of re-engaging and re-integrating those faculties, so that they can be involved in work and not confined to leisure. A real return to nature, to unselfconsciousness, is, precisely, unimaginable, since unselfconsciousness would be unmediated experience.

Campbell's argument is that consumerism is a version of Romanticism because of this positioning of the gazer at the margin, looking in. Because this observer only has brief glimpses and intimations, the idea is established that pleasure is necessarily fleeting. Joy is experienced in sudden revelations; these were the emotions poets would recollect in tranquillity and recreate as similarly intense momentary effects in poetry. Romanticism encouraged 'the sense of a fundamental discrepancy between the satisfactions which life offered and those pleasures which could be enjoyed in imagination' (Campbell, 1987, p. 192). Imagination 'became the most significant and prized of personal qualities' (p. 193), because it was the faculty that could bring these precious glimpses of unalienated life. And what the Romantic imagination became, in its consumerist form, was the day-dreaming imagination. Each of us is 'an artist of the imagination', able to 'create an illusion which is known to be false but felt to be true' (p. 78). The distinctive feature of this new hedonism is that longing is itself a form of delight: 'the desiring mode constitutes a state of enjoyable discomfort, [so] that wanting rather than having is the main focus of pleasure-seeking' (p. 86).

Actually coming into possession of the object of desire brings this day-dreaming phase to an end, but, as the imagined pleasures have the quality of infinity that enables them to exceed those of tangible, time-bound reality, and as imagination is the main recourse of alienated subjectivity, the day-dreaming will soon begin again, transferred to a new object of desire. This is why autonomous imaginative hedonism, unlike traditional hedonism, is not mistrustful of the new, and why modern consumerism is so quick to discard objects that were recently desired:

> The fact that a so-called 'new' product may not, in reality, offer anything resembling either additional utility or a novel experience is largely irrelevant, as all real consumption is a disillusioning experience in any case. What matters is that the presentation of a product as 'new' allows the potential consumer to attach some of his dream pleasure to it, and hence to associate acquisition and use of the object with realisation of the dream.
>
> (p. 89)

Because the essential motivation is in the dreaming and desiring, not the actual possession, Campbell suggests that it is a mistake to denounce consumerism as 'materialistic'. But it has material consequences, which environmentalists see as becoming catastrophic.

Campbell's thesis may be depressing for environmentalists, insofar as it implicates Romantic patterns of desire in consumerism and thus suggests that the Romantic tradition may not after all be the source of radical alternatives it seems. All the versions of 'green' pleasure I have discussed here can be traced to Romantic ideas. But Campbell does see the connection between Romanticism and consumerism as a fundamentally ironic one; an unintended consequence and a Weberian irony of history (see Campbell, 1987, p. 209). It is an irony that might conceivably cut both ways. If Romanticism provides the structure of desire that motivates consumerism, then Romanticism remains powerfully latent in contemporary culture: there to be renewed in non-consumerist forms, if there is sufficient motive. Wendy Wheeler sees Romanticism as having been a 200 year holding operation: vital as the main movement of resistance, but flawed by an excessive preoccupation with the heroic subjectivity of the lonely, sensitive but alienated individual, and by too much readiness to identify all of science with the Cartesian mechanistic tradition. Now Romanticism can be seen as a bridge between pre-industrial sensibility and the post-industrial sensibility that will emerge when the implications of biosemiotics and complexity theory have been so thoroughly digested by culture as to become the new common sense. When will this be? The question is urgent – no other form of literary and cultural criticism labours under such a sense of time-pressure as ecocriticism. I will conclude by saying something about that urgency, and its effect on 'green' pleasure.

The crisis

Most scientists in the relevant fields agree that potentially catastrophic climate change is taking place, and that it is due to increases in atmospheric carbon produced by activities that are the basis of our ordinary lives. The scientific arguments must continue, but when the consequences of ignoring the warnings are likely to be so unimaginably terrible, it is hard to see how anyone can justify the risk of doing nothing. It is one thing to be sceptical, in the sense of remaining open to all arguments and accepting the possibility that catastrophic climate change will not happen. But how can anyone – especially any non-expert – justify feeling so confident it will not happen as to advocate ignoring the preponderance of scientific opinion?

The likelihood, as reported in February 2007 by the Intergovernmental Panel on Climate Change (see Lanchester, 2007, p. 6), is a rise in average global temperatures this coming century of between 2 and 4.5 degrees centigrade. It could be higher: six degrees or more, if feedback effects are severe. Recently, several 'popular science' books have attempted to imagine and explain what this is likely to mean. They include James Lovelock's *The Revenge of Gaia* (2006), Fred Pearce's *The Last Generation* (2006), George Monbiot's *Heat* (2006) and Mark Lynas's *Six Degrees* (2007).

Six degrees, says Lynas, would produce similar ecological conditions to those that caused the end-Permian extinctions, in which about 95% of all species disappeared (see Lynas, 2007, p. 243ff). Dead anoxic oceans, methane fireballs, only the polar regions habitable: this, admittedly, is at the far end of likelihood, as currently judged, but possible. Lovelock reports the opinion of some scientists that 'if global temperatures rise by more than 2.7°C the Greenland glacier will no longer be stable', and that 'a rise in temperature globally of 4°C is enough to destabilize the tropical rainforests' (Lovelock, 2006, p. 51). Both developments would produce huge feedback effects. Pearce asserts that 'we are in all probability the last generation that can rely on anything close to a stable global climate' (Pearce, 2006, p. 293).

Since I am writing about pleasure and consumerism, I want to say a word about how these books make me feel. Reading them gives me a panicky sensation, a sort of danger-tension that does not move to any crisis or release. If one of these futures is true, it seems to follow that the life I know is somehow unreal, a delusion, because it has not equipped me to imagine such a future. But then I look around and touch things, and the life I know has solid surfaces. I cannot fully feel this news to be real, because I am not behaving as if I believed it.

Yet I cannot feel it to be untrue, either, which would release my tension. There is no plausible alternative scenario. So there is an impasse. A reversal has taken place. Rather than acting on the evidence of what I am told on good authority, I am waiting for my actions to tell me that I believe. Reviewing a clutch of these books, the novelist John Lanchester recently expressed a similar idea: 'I suspect we're reluctant to think about it' (he wrote) 'because we're worried that if we start we will have no choice but to think about nothing else... We deeply don't want to believe this story' (Lanchester, 2007, pp. 3, 6).

Since I read it six months ago, this comment has haunted me a little. Lanchester does see the implications of climate change. This is not denial, but a much more contorted emotional manoeuvre. He is saying that he has a premonition of what it would – or will – be like to take climate change seriously: quite a physical premonition, of an unprecedented state of mind. And he cannot do it yet, because he does not have to.

Lanchester can envisage only two options: not thinking about it and thinking about nothing else. Nothing in between seems imaginable. The threat is of such seriousness and probability that the only proportionate response is to be obsessed with it to the exclusion of all else, but this would not be tolerable, so one shields oneself from facing it at all. Considering the emotional burden of the tidings they bear, the climate change books are heroic. Perhaps the cost of this is that their calmness can seem a little eerie, and that the emotional range of the prose is for the most part very limited. Typically, the calmness will be punctured by awkward jokes, strangely unequal to the subject-matter. None of these books is very emotionally literate, in the sense of being able to offer much recognition of the emotional effect they are likely to have on readers, or minister to those emotions.

But those emotions are largely uncharted territory. Slavoj Žižek has suggested that environmental crisis was an instance of the Lacanian 'real', unrepresentable and therefore incommensurate with our understanding, so that no response could be adequate. Right across the range of reactions – denial, complacency, activism, panic – responses will be inauthentic, simulated, self-dramatising space-fillers:

> What is at stake is our most unquestionable presuppositions, the very horizon of our meaning, our everyday understanding of 'nature' as a regular, rhythmic process... Hence our unwillingness to take the

> ecological crisis completely seriously; hence the fact that the typical, predominant reaction to it consists in a variation on the famous disavowal, 'I know very well (that things are deadly serious, that what is at stake is our very survival), but just the same... (I don't really believe it, I'm not really prepared to integrate it into my symbolic universe, and that is why I continue to act as if ecology is of no lasting consequence for my everyday life).'
>
> (Žižek, 1991, p. 34f)

According to this view, we *cannot* respond adequately to the crisis. The scale of the possible catastrophes is incommensurable with any personal narrative we can produce, in any genre. In consequence, the whole notion of narratable experience falls away, making it impossible for these futures to seem real. Making them seem real – finding narrative modes that connect them with our present experience – is the great challenge for cultural environmentalism.

In the meantime, we have something eerily like the Romantic sublime: an experience we can glimpse ahead of us, but that we recoil and fall back from. It is too transformative, too incommensurable: too wild. This chapter has looked, a bit, at how 'wild' can be exciting: how it has long been a powerful – though paradoxical – ingredient of pleasure. Can this new, daunting wildness become so? It looks to be more uncompromisingly real than most of the pleasurable wildnesses we contrive.

Concluding reflections

Some leading campaigners for action on climate change have recently gone out of their way to talk about environmentalist exhilaration and joy. Bill McKibben, in a public lecture of summer 2007, described the present historical moment as astonishingly perilous but also as 'a beautiful moment', because of what he thinks it will make us discover in ourselves. Al Gore declares a similar hope. In his most recent slide-show sequel to *An Inconvenient Truth*, he says that 'we ought to approach this challenge with a profound sense of joy and gratitude', because we have the opportunity to become 'another hero generation', comparable to the American generations that won independence, abolished slavery, fought Nazism and campaigned for civil rights (Gore, 2008). Gore invokes a specifically American patriotic mythology here, but there is a British equivalent. James Lovelock and George Monbiot have both compared the collective spirit called for by the new crisis with the fabled British communal spirit of the Second World War: the 'Dunkirk spirit' or 'Blitz spirit'.

Such appeals are very culture-specific and run the risk of alienating those who do not identify with the myth invoked. They also risk collusion with nostalgia, coercive sentimentality and even militarism. Their primary motivation, in this perilous moment, is probably to find a counter-narrative to that of overawed pessimism, and one that can play consumerism at its own game by tapping into popular collective fantasy. The pleasure they hold out is only hedonistic in a very paradoxical way: the idea, traditional enough, is that what makes us come alive, in the sense of discovering our real needs and engaging our full senses, is a danger big enough, obvious enough and immediate enough to call us to those senses. From this point of view, environmentalism's big problem is how to reveal the danger to the popular imagination in a way that makes it immediate enough to alert our bodily senses, since we do not seem to be constituted, either by culture or by evolution, to respond existentially to threats that will materialise in 20 or 30 years' time, however serious they are. Lovelock, in his most recent statements, seems resigned to the impossibility of this. 'Enjoy life while you can', he is reported as saying. '[I]f you're lucky it's going to be 20 years before it hits' (Guardian, 2008). This is a kind of call to hedonism, but a despairing one, born of the conviction that nothing will make us respond until the disaster is a present catastrophe. Since we can do nothing for now, we may as well try to forget, if we can.

Evolutionary psychology offers one explanation for this impasse, though it is one that the academic Humanities have only recently, and with great reluctance, begun to explore. Broadly, the idea is that our tastes and impulses are distantly derived from hunter-gatherer life in the Pleistocene era, and that nothing in our evolutionary development has prepared us for the demands of the threat from climate change, a threat that remains primarily theoretical rather than registering upon our physical senses and thus motivating our fight or flight responses. What we have is a new version of the break between thinking and feeling regretted by Coleridge in 'Dejection: An Ode' (1802):

> Yon crescent Moon, as fixed as if it grew
> In its own cloudless, starless lake of blue;
> I see them all so excellently fair,
> I see, not feel how beautiful they are!
>
> (Coleridge, 1997, p. 308)

We know, not feel, how dangerous climate change is. That is, we do not feel it as a natural animal feels on its senses a tangible threat that makes its body leap into action. Instead, in our consumerist appetites, we are like the fox in the henhouse killing every bird in sight because his evolved impulse is simply to pounce on a fluttering bird. In deep evolutionary history the fox did not encounter masses of birds unable to fly away. Nothing prepared his impulses for the henhouse. In a similar way, perhaps, people find it hard to stop driving because evolution disposes us to sensuous enjoyment of speed, or find it hard to stop eating fat, salt and sugar because our appetites evolved when these were hard to obtain. What the crisis requires of us is counter-impulsive.

Our evolved senses and impulses, then, are part of the problem. David Abram's enraptured account of the reawakening of all our natural sensuous alertness should be seen in this context. People conditioned by industrial consumerism, he suggests, need to retrain those senses, tearing them away from the false, environmentally disconnected gratification that is consumerist hedonism. Abram identifies this re-embedding of the senses as the true antidote to consumerism; much better than simply telling people to desist from their harmful pleasures. The re-embedding that he recommends would return our senses to something close to the intensity they had when it was necessary, in hunter-gatherer life, to be wholly and constantly alert to the stimuli of the moment – except that he wants this without the precariousness and shortness of the life in which these senses evolved. He wants it, therefore, as a kind of ecstatic pleasure, and a flood of generous and perceptive sympathy with the human and non-human world, not as a set of constantly needed survival mechanisms. *Ultimately*, however, the new, or renewed, sensuous awareness he desires would indeed be a survival mechanism, since he sees it as our main chance of bringing ourselves to make the changes needed to save us from catastrophic climate change.

I think it is inadequate, therefore, to dismiss the contradiction in Abram's desire as a piece of bad faith in the 'noble savage' tradition – a desire for a leisurely simulacrum of nature rather than real Hobbesian nature. It may be that the contradiction is not Abram's but the world's. We need two things powerfully at odds with each other. One is the ability to lose ourselves in the moment and in the task, freeing ourselves from the insatiable restlessness of what Colin Campbell calls 'self-illusory hedonism'. The other is the ability to look far ahead, take the long-range view that science offers us, and respond to a threat that is not presently registering on our senses, at least not sufficiently to motivate change. We need to be able to plunge more deeply in – to friendship, love, parenting, craftsman-like work, arts, bodily sensations, caring for people and things rather than discarding them, pleasure in the impulses of others, pleasure in the seasons and the way they return and measure out life – in accordance with the principles of 'slow food'. Yet we also need to be able to surface swiftly and look further. Environmentalists are simultaneously working for deep changes and feeling that there may not be time for those changes, and that the first need is for emergency measures to enable us to survive the imminent catastrophe. Lovelock, in *The Revenge of Gaia*, sees the need for both strategies. He looks to Deep Ecology for the fundamental reorientation of our pleasures and values that will enable us to live sustainably in the long run, but also to the immediate preparation of technological solutions – pre-eminently nuclear power – to buy time in which these deeper changes might have a chance to take hold. How to develop both abilities is the question: how to make them willingly answerable to each other, and prevent them from negating each other.

Romanticism shows us something about this, in the ambivalence, the having-it-both-ways, of its joyful rediscovery of nature. The bad faith to which Romantic desire is prone is that it longs for a life into which it does not dare to leap, and comes to savour the longing, the prospect, the tourism, the edge of unexplored experience, as an *alternative* to immersion. This is the pleasure Colin Campbell sees as an essential ingredient of consumerist hedonism. The Romantic sense of infinite possibility seems to exceed what absorption in any particular experience could provide, and thus keeps the Romantic self in a condition of alienation, positive in its visionary joy but negative in its insuperable sense of exile. Romanticism loves the idea of going in deep, but continually pulls us out again. To turn this contradiction into a virtuous acknowledgement of both needs is a daunting task, but it is what environmentalism requires. Local and global must both be held in view. Immersion in experience and self-conscious joy at the prospect are both necessary. Cultivating them as distinctive pleasures, freshened by the contrast, can only help.

References

Abram, David (1997; first published 1996) *The Spell of the Sensuous* New York: Vintage.

Campbell, Colin (2005; first published 1987) *The Romantic Ethic and the Spirit of Modern Consumerism* Great Britain: Alcuin Academics.

Coleridge, Samuel Taylor (1997) 'The Eolian Harp', 'Frost at Midnight', 'The Rime of the Ancyent Marinere' and 'Dejection: An Ode', in Keach, William (ed.) *Samuel Taylor Coleridge: The Complete Poems* London: Penguin.

Deakin, Roger (2000; first published 1999) *Waterlog* London: Vintage.

Gore, Al (2008) 'New Thinking on the Climate Crisis' March 2008: see http://www.ted.com/index.php/talks/view/id/243 (downloaded 14 April 2008).

Griffiths, Jay (2006) *Wild* New York: Tarcher Penguin.

Guardian (2008) 'Interview with James Lovelock', *The Guardian* 1 March 2008, p. 33: see http://www.guardian.co.uk/theguardian/2008/mar/01/scienceof climatechange.climatechange.

Harvey, David (1996) *Justice, Nature and the Geography of Difference* Oxford: Blackwell.

Hegland, Jean (1996) *Into the Forest* New York: Bantam Books.

Ingold, Tim (2000) *The Perception of the Environment* London: Routledge.

Lanchester, John (2007) 'Warmer, Warmer' *London Review of Books* 22 March 2007, pp. 3–9.

Louv, Richard (2005) *Last Child in the Woods: Saving Our Children from Nature-Deficit Disorder* Chapel Hill: Algonquin Books.

Lovelock, James (2006) *The Revenge of Gaia* London: Allen Lane.

Lynas, Mark (2007) *Six Degrees: Our Future on a Hotter Planet* London: Fourth Estate.

Matless, David (1998) *Landscape and Englishness* London: Reaktion.

McKibben, Bill (1990) *The End of Nature* London: Viking.

Monbiot, George (2006) *Heat* London: Allen Lane.

Pearce, Fred (2006) *The Last Generation* London: Transworld.

Ross, Andrew (1994) *The Chicago Gangster Theory of Life* London: Verso.

Soper, Kate (1995) *What is Nature?* Oxford: Blackwell.

Wheeler, Wendy (2006) *The Whole Creature* London: Lawrence and Wishart.

Wordsworth, William (2004) 'Expostulation and Reply' and 'The Prelude', in Gill, Stephen (ed.) *William Wordsworth: Selected Poems* London: Penguin.

Žižek, Slavoj (1997; first published 1991) *Looking Awry: An Introduction to Jacques Lacan through Popular Culture* Cambridge, MA: MIT Press.

Part III

Everyday Consumption

8
Happiness and the Consumption of Mobility

Juliet Solomon

I must start by saying that in many ways this chapter explores issues which have long been avoided by most of those who purport to be concerned with transport issues probably because of their emotional and spiritual associations. I went into the transport research field essentially because I am an environmentalist who is concerned about the negative physical environmental impacts of the car, which were emphasised in the 1994 Report of the Royal Commission on Environmental Pollution (RCEP), and because I find the topic of mobility fascinating, and hoped to learn more. But pretty quickly I realised that the members of the transport profession, based as it is in engineering disciplines, may understand how to build a road, but do not understand society's outlook on mobility – why we 'whizz about' so much. In 1950 the average Briton travelled approximately five miles a day. Now it is closer to 28 miles a day, and forecast to double by 2025 (Adams, 1999). Globally, this represents a massive increase.

The assumption made in 99% of the transport literature (even the so-called 'motivational' literature – see, for example, Jones' (1983) *Understanding Travel Behaviour*), and from which most transport policies are derived, is that travel and transport constitute what economists term a *derived demand* and are therefore not consumed for their own sake. People only go from A to B because they want to do activity C or purchase commodity D; the actual 'going' is rather a nuisance. Although a few voices (Mokhtarian and Salamon, 2001; Solomon, 1998) have challenged this assumption, it is only very recently that any discussion of the implications has featured in the transport literature and debates.

The car as commodity features prominently in the discourses about the increasing consumption of material goods (many of them ephemeral) that have been central to the consumerism debate. The notion that 'I shop, therefore I am' is increasingly embedded in the culture (Benson, 2000). Shopping gives meaning to life. 'Retail therapy', the purchasing of material goods, provides comfort. The car's iconography and symbolic role as a consumer good have been quite comprehensively written about (Carrabine and Longhurst, 2002; Marsh and Collett, 1986). Cars are talked about, cars are evaluated and displayed routinely in news media (BBC, 2006a). The motoring supplements of some of the Sunday newspapers are as fat as the business supplements. The material aspect of travel in the shape of the car and accessories to the car gets a good airing. However, beyond a general celebration of the speed and handling qualities of the vehicle, all this attention to the car as commodity contributes little to our understanding of why travel and movement have become so important to so many people.

For their part, critics and opponents of the consumer society mention travel as an add-on to their arguments about how to persuade people to purchase less (or purchase differently), apparently not regarding travel/movement as a consumer good distinguishable from material goods in general. The following (from the 'Responsible Consumerism' website) is typical of much of the thinking:

> The real challenge is to promote more thoughtful forms of consumerism – satisfying needs with minimal use of resources, and at a wider social level, reducing the need for travel and energy consumption (this should also be done at a personal level!)
>
> (Responsible Consumerism, 2003)

Considerations of why this approach might in some respects be oversimple appear, from their absence, not to be taken very seriously. In terms of damage done, social and environmental (and possibly psychological, although that is largely, so far, an untold story), but also in terms of human pleasure and self-fulfilment, the consumption of movement is a social phenomenon of which the counter-consumerist movement should have a meaningful and complex analysis. Here, I want to explore the extent to which movement, even commuting, is a significant part of the spectrum of human pleasures, essential to individual identity formation. Unless we understand this, we will not comprehend what might be the implications for human happiness of a reduced consumption of mobility.

The primacy of movement

> Movement is one of the great laws of life. It is the primary medium of our aliveness, the flow of energy going on in us like a river all the time, awake or asleep, twenty-four hours a day. Our movement is our behaviour; there is a direct connection between what we are like and how we move.
>
> (Whitehouse *et al.*, 1999, p. 17)

Movement is obviously fundamental to the human condition. The absence of movement is death. The kinds of travel that many citizens of wealthy countries have nowadays become accustomed to are integral to contemporary lifestyles: in the immediate aftermath of 9/11, President Bush encouraged people to continue flying, and his message was echoed by the British Prime Minister, Tony Blair, who urged the British people 'to work, live, travel, and shop' as before – rather than surf the Internet and start spending (BBC, 2006b). Here, travel is not seen as an unwanted auxiliary to a more important activity (shopping); it is movement itself that we need to consume, even if we do not quite realise it.

Even columnists who may present themselves as anti-materialistic, including some writing in the national newspapers, seem sometimes to be commenting from the comfort of their Volvos on the size and nature of other people's cars – if those cars happen to be 4x4s, which seem to be regarded as a legitimate target for wrath and scorn. (In 2005, it is worth noting, 187,000 of these were sold in the UK; a great many people evidently believe them a necessary or at least desirable possession (BBC, 2006a,b).) Equally, people who regularly take what some of us might view as unnecessary work journeys may happily deride those who flit off for a break for no particular reason but that an air fare is cheap and they think there might be some benefit in 'getting away' (RSA, 2006).

However, the extent of mechanised travel, and its increasing embeddedness in people's lives and routines, is undoubtedly causing enormous environmental and social problems, even if we believe and hope that it is also expanding human possibility and outlook. If we want to try to limit the damage while trying to hold onto some of the benefits, we need an analysis which understands and acknowledges what the movement is contributing to our existence. Unless the technofixers can, literally, produce miracles, it will not be long before we are to some extent forcibly grounded by a shortage of fuel. If we are not to be deprived by the concomitant changes, will have to find other ways of being that can substitute for the mechanised movement. In this chapter, I aim to suggest and discuss some aspects of mobility as a non-material good. Of those human needs and qualities whose sustenance or development especially depend on movement, I am going to consider just three. These are connected with the role of *physical sensation*; the need for what is now called *personal space*; and the need for *hope*.

Sensation

> I sprang to the stirrup, and Joris, and he;
> I galloped, Dirck galloped, we galloped all three;
> "Good speed!" cried the watch, as the gate-bolts undrew;
> "Speed!" echoed the wall to us galloping through;
> Behind shut the postern, the lights sank to rest,
> And into the midnight we galloped abreast.

Robert Browning's well-known poem 'How they brought the good news from Ghent to Aix' (Browning, 1890) tells the fictional story of three messengers who rode at the fastest pace their horses could go in order to deliver news that would end a war. The poem is an invigorating and life-enhancing celebration of movement, speed and energy (although 'speed kills', even in Browning: the efforts made by two of the messengers' horses to maximise their speed result, tragically, in their collapse and death). Just reading this poem, especially aloud, can produce an adrenalin rush, even though we neither know nor much care what news the messengers were carrying, or to whom, or why. What grips us has nothing to do with our intellect or even our ear, but everything to do with our emotions. During this extraordinary ride, Browning manages to pull us through some of the highs and lows of being – anxiety, joy and fear – and to make us feel excited anticipation, as well as a sense of threat and, at the same time, physical stimulation. Finally, to our relief, he allows us a time of repose when the ride is over.

Anybody who has ever ridden on a roller coaster (and if you are not one of those people, it is well worth a try) will know how reminiscent the poem is of that experience. The roller coaster is about nothing except subjection to the sensation of movement. It does not matter where it is, and it does not matter where it goes. What matter are the changes in consciousness which it triggers, as you are turned round about, shot into the air, thrust into the depths, suspended perhaps upside down, with your views inverted and your horizons blurred. For those who like it, it is enjoyable and, in a way, addictive, the kind of experience which some might call a 'serotonin fix'.

However, for the 'fix' of movement to be effective it need not be fast or masquerade as dangerous (the roller coaster is in fact totally safety-proofed). At the other end of the spectrum is the pleasure derived from riding on a canal boat, a sensation enjoyed by increasing numbers taking their holidays on barges. For L. T. C Rolt, who was fascinated by the effect that a variety of different types of movement had on him, even the movement of the calmly ambling narrow boat, which he described as 'no motion, or ... motion asleep', becomes addictive (Rolt, 1946, p. 57). He contrasts the 'placid three-miles-an-hour gait' with the traffic driving by on the road above one of the canal bridges to show just how different the two types of movement make him feel: 'The rush of traffic on the road above seems to become the purposeless scurrying of an overturned ant-hill beside the unruffled calm of the water, which even the slow passage of the boats does not disturb' (p. 12).

The experiences of the reader of the Browning poem, the roller-coaster rider and the canal-boat traveller may not appear to have much in common. What unites them is that in all three cases, the pleasure derived from them has its origins in the sensations (real or imagined) of physical movement. These sensations tend to become a kind of fix, consumed in increasing quantities – as is the movement of commuters wending their daily ways to work.

The pleasures involved here evidently have both physiological and psychological aspects and effects. Isolating and separating these is bound to be difficult, as both physiology and psychology are involved in the ideas of horizon, potency, novelty and energy. Only movement extends a person's knowledge of distance and place and familiarises them with spatial and experiential horizons of which they may previously not have been aware of. However lumbering the movement, even a dreary and slow bus ride to work or an hour spent in a traffic jam in the car cannot be totally disconnected from all the wonder of new awakenings.

Psychophysiology's influence as a discipline seems to be unjustly dwarfed by other branches of both physiology and psychology. Intuitively, however, it seems obvious that a person's movements and their feeling states must be connected, and that movement can also be used deliberately to trigger changes in mood. The philosopher Maurice Merleau-Ponty (1962) placed the body at the centre of his thinking. The body was to him a source of understanding: all experiences came through the body and involved movement. Movement was therefore fundamental, and Merleau-Ponty argued that if it was treated merely as a 'handmaid of consciousness', rather than as intrinsically significant, this was a misrepresentation and a barrier to understanding. He insisted that feelings which can be experienced while moving cannot be experienced in other ways.

A number of writers, from as early as the sixteenth century, have made similar observations. They range from Robert Burton, who attributes some forms of melancholy to an absence of movement (Burton, 1621, p. 242), to Bruce Chatwin (1996), who believed that the straight legs and the striding walk of the human being were accompanied by a migratory drive to walk long distances in the appropriate season, which needed satisfying. If not satisfied, according to Chatwin, the drive to move found outlets in violence, greed, status-seeking or a mania for the new – all of which, unlike long-distance walks, easily result in confrontation, frustration and destructiveness, as well as excessive and unnecessary consumption. In other words, unless we walk we cannot be constructively and totally human; and we are likely to become, to some extent, antisocial. Sadly, Chatwin's early death prevented his development of this thesis.

Our first experience of movement starts passively, and then becomes active, months before birth. It is the movement of the embryo that effectively 'introduces' the developing foetus to its mother and signals that it is alive. This passive in-womb rocking experience of the foetus is widely believed to be a state of innocence and bliss. Combining, as it does, movement, warmth and safety, experience suggests that there is every reason to share the belief.

It is very striking how much enjoyment children, as they learn to move independently, can be observed to take from their capacity for locomotion. Crawling and walking is just the beginning; the manifest thrill the child gets from his first walk is, however, put well into the shade when compared to the thrill that child exhibits when she has just learned to ride a bicycle. Watching a child who has just acquired the skill steam across a playground or down a path in a park, enjoying the enormous difference that the machine makes to their self-powered capacity for speed and mobility, is an experience that takes many adults back to the feelings of achievement and liberation we experienced when doing the same many years ago.

The exhilaration experienced during active movements powered by human energy like riding a bicycle or roller skating can be found in other movements like skiing, horse riding, certain types of dancing, rollerblading, rowing or canoeing, riding the travelator, and possibly even swimming. All of these, with the exception of dancing and swimming, use add-on extensions to enlarge human capacities, but rely on human motive power. From an early stage of our lives, we move for movement's sake and enjoy the experience of movement. But as we grow up, gradually almost all our movements seem to become instrumental. Almost all of them are likely to become mechanised. We walk to get to the shop, although increasingly we drive; but we habitually drive to go on holiday, we fly for a business trip, we take a bus to the town centre (Department for Transport, 2006). We cease to enjoy the experience or even to notice that we are actually moving unless we skate, dance, bike, practice Tai Ch'i or take part in other formalised movement activities including walking. In all too many societies – but thankfully not yet all – by the time we are adults, we have ceased to allow ourselves to indulge in spontaneous creative expression in physical movement, except for some forms of dancing in private places (Londondance.com), or 'going for a drive'.

In so doing, we are cutting ourselves off from a rewarding part of human experience and the concomitant extension of our emotional and spiritual vocabulary. The experience of a walk, or of running, or of flying in a biplane, of sitting on a train, or of driving a car all give rise to very different feelings. Many of these emotions will be new to the person feeling them and will only have surfaced in their emotional repertoire because of the way in which they are moving – or being moved. To deny the possibility of such movement or to ignore its impact is to cut off these dimensions of inner feeling.

> The experience of new quality of movement provides a change in feeling, another dimension of the self. Change can also take place through the discovery of previously unavailable feelings. For the ones who move like bulldozers, the longing to experience themselves as yielding and delicate results in a changed quality of movement. So movement can lead to new feeling; feeling can lead to new movement. They are, in some way, one.
>
> (Whitehouse *et al.*, 1999, p. 60)

The connection between our state of being and feeling and our locomotion applies not only to our own unaided physical movement, but to aided movement as well. All tools, from knives and guns to roller skates, cars and travelators, extend our motor capacities and effectively become a part of ourselves (the word 'exosomatic' was invented to refer to those instruments which, though not parts of the body, are nevertheless functionally integrated into ourselves). Thus the significance of 'how we move' is not confined to our bodily movements (Medawar, 1982, p. 185) and cannot necessarily be separated out from 'how we are moved' (That is, by a machine). In some cases we may well feel quite bonded with the machine or accessory, in the same way as a rider does with his horse. Writer and pilot Antoine de Saint-Exupéry had this kind of relationship with his aeroplane (Saint-Exupéry, 1940, p. 72f).

Mihaly Csikszentmihalyi, in *Beyond Boredom and Anxiety*, studies activities which he describes as 'flow' activities, activities which appear to provide great pleasure to those undertaking them. When in 'flow', the subject has reached the point of being able to resonate his abilities with the surroundings, whatever they are, and 'is in harmony with the world. He can be in solitary confinement or in a boring job ... but ... he will still be enjoying himself' (Csikszentmihalyi, 1975, p. 206). Csikszentmihalyi studied people engaged in absorbing activities like rock climbing and chess playing. He discovered that when they were involved in these activities they were able to attain a deeply satisfying quality of subjective experience, and a loss of self-awareness very similar to the feelings experienced by the roller-coaster rider or the biker, who were 'at one' with a universe outside themselves, or indeed the walker whose walk became almost a meditation. This is 'flow'. The kind of reaction he got from those questioned was that their enjoyment included forgetting the passage of time, freedom from boredom and worry, and often control and the use of talents and skills. Sports, sex, and driving (at their best) are examples of flow activities; so is rhythmic walking, or running.

The concept of Peak Experience has some similarities to that of 'flow', but goes well beyond it, and is familiar from the work of Abraham Maslow, and, unsurprisingly to those who are familiar with him, Colin Wilson. The peak experience, which is particularly common among the religious, people whose lives involve making discoveries, and sportspeople, is one of life's great enhancers.

> Feelings of limitless horizons opening up to the vision, the feeling of being simultaneously more powerful and also more helpless than one ever was before, the feeling of ecstasy and wonder and awe, the loss of placement in time and space with, finally, the conviction that something extremely important and valuable had happened, so that the subject was to some extent transformed and strengthened even in his daily life by such experiences.
>
> (Maslow, 1970, p. 164)

The feeling described here is not necessarily associated with movement, but it can be and often is. The roller coaster would be an example, as would Schivelbusch's accounts of the feelings engendered by the first rides on early trains, when people who had never even contemplated speeds faster than a horse suddenly found themselves being transported at what seemed to them a shocking and awesome 30 miles an hour (Schivelbusch, 1980, Chapter 2). Some runners say they experience this: it is commonly known as 'the runner's high' (caused by the release of endorphins after running).

Peak experiences can be compared to some varieties of religious experience, in that they tend to be transcendental, and often give a sense of purpose to the individual, a sense of integration. Maslow claims that peak experiences can be therapeutic, as they tend to increase free will, self-determination, creativity and empathy. He believes, therefore, that we should study and cultivate peak experience, so that we can teach them to those who have never had them or who repress or suppress them, providing a route to achieve personal growth, integration, and fulfillment.

Reviewing the forms of movement I have been discussing here, and taking into account which of them consumes more resources and does more damage, we are likely to conclude that some of them – especially driving and flying – are becoming environmentally unaffordable. We are also likely to note that other forms of movement – Rolt's canal-boat journeys; walking; cycling – might be enjoyed more often, by more people, in another kind of society, one in which high speed was not a priority. Insofar as we can, we need to satisfy the psycho-physiological need for sensation derived from movement through these slower, human-powered kinds of travel. At present, these gentler ways of moving about are marginalised, and made difficult and dangerous, by the dominance of mechanised transport.

Life space

Over 40 years ago, sociologist Sebastian De Grazia calculated that most Americans spent, on average, between 10 and 20% of their working life going back and forth between home and work (De Grazia, 1962, Chapter 1). He undertook various surveys to try to establish what the journey meant to people, and he found a great number who enjoyed it. As an example, he cites 'the man who rides the 8.05 and gets into New York at 8.55'. This man, who (crucially) is able to get a seat on the train, claims that the ride to and from work is actually one of the high spots of the day, it being the only time available to him for nearly an hour of relaxing reading or, effectively, meditation. De Grazia found people who would use the time to read over the business or financial sections of the paper on the ride to work, and perhaps to go over office papers on the ride home: people who, equipped with computers and mobile phones, would now be making use of the train as office (Holley, Lyons and Jain, 2006). For those in cars, the time between home and work may be the only chance they have in the whole of their lives for solitude. A study by Koslowsky, Kluger and Reich (1995) had findings for drivers similar to those of de Grazia for rail commuters. Many solo drivers said that the time spent in the car on the way to work or back home was the only time when they were able to think and plan their future activities. Many commuters may perceive such time on the journey home as a chance for peace and quiet after a hectic day; it is the time for unwinding before facing the family or some other activity that will require the commuter's active participation.

It could not be said of either of these groups of commuters that the movement that accompanies their quiet times is essential for the fulfilment of the space-seeking purpose – meditation, or a church service, might do as well. However, while in motion, the traveller is distanced from contact with the earth, and is in a bounded, safe place in which he is in a state of literal and metaphorical suspension, as if in a chapel for morning prayer. The movement of the train or the car may be soothing, like the rocking in the womb and cradle, with a moving picture always available through the window. Here he is temporarily freed from everything that ties him to earthbound responsibility and the realities of permanence – or can be, if he has switched his phone off: if not, he misses out on one of the main benefits of being on the move. The world he inhabits on the journey is a kind of dream-world, real yet not real, sometimes described as liminal. At his destination, if getting off the train, he may well become part of the flow of moving workers, travelling together and becoming part of a greater purpose. And being part of the common purpose might be seen as legitimising the consumption of a journey which may be replacing many of the functions of a local, familiar place of communal worship.

Hope, rest and movement

It is both possible and profitable (though not in a financial sense) to view the whole of a person's life through the dialectic of movement and rest, inside and outside, dwelling and journey. Changes in place – from hour to hour, day to day, year to year, early adulthood to middle age – can all be interpreted in terms of a need to move and rest, to stay in a particular place for a time and then move elsewhere. These temporal thresholds correspond, at least in part, to the changing needs and feelings of the subject, who after a certain time begins to feel discomfort, boredom, wanderlust, or some similar emotional push or pull which moves him or her to another place or situations (Seamon, 1979).

The 'round world of dwelling' offers a cyclical time, that is, the recurring ties of seasons, or the cycles of birth and death, of planting and harvesting, of meeting and meeting again, of doing and doing over again. It offers a succession of crops, of duties, of meals, of sleeping and waking, forever appearing and reappearing. It offers a place where fragile objects and creatures can be tended and cared for through constant, gentle recurring contacts. But it does not take the individual forward.

Movement, in contrast, as we have seen, with its links with horizon, reach, strangeness and difference, helps the person to assimilate places and situations into his world of familiarity. In this sense, movement widens the sphere of at-homeness and dwelling, and expands the personal territory. Just as importantly, it provides hope, the prospect of change from a possibly unsatisfactory resting place or time, the possibility of uncertainty, and, associated with leaving the unsatisfactory resting place, a sense of purpose. This is true whether we consider the daily or thrice-daily journey or, at the other extreme, the world tour. Man is 'on the run' as never before, even though electronic communications make this increasing mobility unnecessary (Cairncross, 1997). One has to ask oneself what this says about how much hope or purpose he has when not in motion.

It is quite instructive, in this context, to look at what people do with their time when transport systems are speeded up and they have the chance to shorten their journey. If travel and movement served no purpose but to get from home to a particular destination, they would take advantage of the shorter journeys. But they do not. When journeys are speeded up by the construction of a new road, or a new railway line (such as the fast RER line which goes all the way across Paris), people use the faster speeds to extend their range of destination choice rather than to save time from the journey. There seems to be an approximate maximum amount of time they are prepared to spend travelling to work for any given destination: to some extent this is a matter of social convention, but it might also be a matter of desire, although the traveller may indulge in a few ritual grumbles about it. And if the people you know take an hour to get to work, then you are not abnormal if you also take an hour.

The time seems also to be related to the destination and the status involved in being there. For example, people getting into London, as H. G. Wells pointed out in 1902, had over the centuries been prepared to take about an hour: this was the time it once took to walk about 4 miles, then to ride 8 miles, and by 1902 to travel 40 or 50 miles (now 150 miles) by train. People commute from Brussels to the City of London in not much more time, now that there is an airport near their London office. However, this does not apply to the poor. Anyone living on the minimum wage is, very reasonably, quite likely to regard any work journey over 20 minutes as unacceptable (DETR, 2000). The territory of the poor, the 'waiting classes' who are increasingly left behind in the movement game, is not therefore generally much extended even by such mundane daily movements as commuting.

This perspective on the relationship between where people live and where they work, and on how this changes as fast transport networks develop, suggests significant questions about how people feel about their immediate locality and about home. These in turn have implications for the extent to which it might be possible to decrease the consumption of mobility. If status is related to territorial 'ownership' and long journeys facilitate this ownership, then the journeys will continue, and probably become ever longer. But it is now clear that the long-term environmental costs of superfluous journeys are higher than our planet and our ecosystem can tolerate, beyond the very short run; one way or another, they are going to have to be curbed (RCEP, 1994).

Concluding reflections

The role of mobility in the destruction both of the ecosystem and of the peaceful coexistence of oil-consuming nations with oil producers is becoming increasingly clear even to those who deny that there is an 'environmental problem'. Within the green transport movement, there are various attitudes taken to ways of reducing the use of cars and aeroplanes. The first of these treats current mobility behaviour almost as a disease, sometimes called 'hypermobility', with its overtones of excess. Probably (in this view) nothing short of rationing, high taxes or prohibition will be needed to stop people travelling so much. The second approach, which is concerned with daily mobility, aims to try to alter the built environment to reduce the 'need' to travel, the assumption being that if there is a 'necessary' destination such as a school, a job, or a variety of shops within walking or cycling distance, then people will use these rather than journey to a more distant location, and in so doing will change their travel mode to something less unsustainable than the car.

Both these approaches reflect the commonly held view that travel itself provides no benefits to the travellers; and that what is required is thus merely a substitute for the unsustainable means of movement from location to location. I have argued that this is far too limited a view, and that in many ways, the 'need' to travel is a deeper human need than can be satisfied by taking a bus or train from A to B. If we want to reduce that need, we must establish precisely what pleasures it provides and must design sustainable alternatives.

Movement and mobility as generators of flow experiences, as creators of personal space, as suppliers of hope, as expanders of our world-view: these are important functions which need to exist, and to be convincingly expressed, in all our personal narratives. I have tried here to argue for these considerations to be taken into account when we are trying to persuade people to travel more sustainably. If, as we are just beginning to recognise, mobility, which may be life-saving and soul-enhancing, is now too 'troubled a pleasure' (Soper, 1990) to tolerate at current levels, then in reducing and controlling it we must make sure that we do not lose sight of the larger picture and end up diminishing our humanity.

References

Adams, J. (1999) *The Social Implications of Hypermobility*. OECD report: ENV/EPOC/PPC/T(99)3/Final – available free from the OECD Environment Directorate.

BBC (2006a) see http://news.bbc.co.uk/1/hi/magazine/4829628.stm.

BBC (2006b) see http://news.bbc.co.uk/1/hi/talking_point/3850225.stm *Should urban 4x4 drivers be penalised?*

Benson, A. L. (ed.) (2000) *I Shop, Therefore I Am* New Jersey: Northvale.

Browning, Robert (1890) 'How they brought the good news from Ghent to Aix', in *Selections from the Poetical Works of Robert Browning: First Series* London: Smith, Elder and Co.

Burton, Richard (1977; first published 1621) *The Anatomy of Melancholy* London: Dent.

Cairncross, Frances (1997) *The Death of Distance* (2nd edition 2001) London: Texere.

Carrabine, E. and Longhurst, B. (2002) 'Consuming the car: Anticipation, use and meaning in contemporary youth culture' *Sociological Review* 50(2), pp. 181–196.

Chatwin, B. (eds Born, J. and Graves, M.) (1996) *Anatomy of Restlessness: Uncollected Writings* London: Cape.

Csikszentmihalyi, M. (1975) *Beyond Boredom and Anxiety* San Francisco: Jossey-Bass.

De Grazia, Sebastian (1962) *Of Time, Work and Leisure* New York: Twentieth Century Fund.
Department for Transport (2006) *Focus on Personal Travel* (2005 edition) http://www.dft.gov.uk/pgr/statistics/datatablespublications/personal/focuspt/2005/.
Department of the Environment, Transport and the Regions (2000) *Social Exclusion and the Provision and Availability of Public Transport.*
Holley, D., Lyons, G. and Jain, J. (2006) *Towards an Understanding of the Use and Value of Business Travel Time* UTSG: Dublin.
Jones, P. (1983) *Understanding Travel Behaviour* Aldershot: Gower.
Koslowsky, M., Kluger, A. N. and Reich, M. (1995) *Commuting Stress: Causes, Effects, and Methods of Coping* New York and London: Plenum Press.
Londondance.com Website; last accessed March 2007.
Marsh, P. and Collett, P. (1986) *Driving Passion: The Psychology of the Car* London: Cape.
Maslow, Abraham (1970) 'Religious aspects of peak-experiences', in Maslow, *Personality and Religion* New York: Harper and Row.
Medawar, P. (1982) *Pluto's Republic* Oxford: Oxford University Press.
Merleau-Ponty, Maurice (transl. C. Smith) (1986; first published 1962) *Phenomenology of Perception* London: Routledge and Kegan Paul.
Mokhtarian, P. L. and Salamon, I. (2001) 'How derived is the demand for travel? Some conceptual and measurement considerations' *Transportation Research Park A* 35, pp. 695–719, Institute of Transportation Studies, University of California, Davis.
Responsible Consumerism (2003) see http://www.recycle.mcmail.com/respons.htm.
Rolt, L. T. C. (1946) *Narrow Boat* London: Eyre and Spottiswoode.
RCEP (Royal Commission on Environmental Pollution) (1994) *Eighteenth Report Transport and the Environment* HMSO Cm 2674: ISBN 0 10 126742 8.
RSA (2006) see http://www.rsa.org.uk/journal/article.asp?articleID=755.
Saint-Exupéry, A. (transl. Lewis Galantière) (1966; first published 1940) *Terre des Hommes (Wind, Sun and Stars)* Harmondsworth: Penguin.
Schivelbusch, W. (1980) *Railway Journey: Trains and Travel in the Nineteenth Century* Oxford: Blackwell.
Seamon, D. (1979) *A Geography of Everyday Life* London: Croom Helm.
Solomon, Juliet (1998) 'Reaching hearts and minds' *Proceedings of Seminar C, European Transport Conference,* London: PTRC, pp. 41–52.
Soper, Kate (1990) *Troubled Pleasures* London: Verso.
Wells, H. G. (1902) *Anticipations of the Reaction of Mechanical and Scientific Progress Upon Human Life and Thought* London: Chapman and Hall.
Whitehouse, M. S., Adler, J. and Chodorow, J. (eds) (1999) *Authentic Movement* London: Jessica Kingsley.

9
Gendering Anti-Consumerism: Alternative Genealogies, Consumer Whores and the Role of *Ressentiment*

Jo Littler

In the summer of 2004, during my first visit to New York, I walked past a branch of the world's largest chain of coffee shops, Starbucks. Outside, I saw a young, slightly punky woman sitting on a sheet selling badges. Many depicted a reworked Starbucks logo. On them, the words 'Consumer Whore' replaced the usual 'Starbucks' lettering, and the female Starbucks icon (a mermaid commonly known as the 'Siren') was shown wearing a crown with a dollar logo, gazing ahead vacuously, holding a Starbucks coffee in one hand and a mobile phone in the other. I liked it. I bought one.

The badge seemed interesting to me on several levels. Most obviously, it was hostile to Starbucks. Now, clearly, we might do a lot of nuanced readings of the more positive cultural features Starbucks offers, such as expanding social spaces for conviviality, cosmopolitanism, and a relative democratisation of 'luxury taste' sensations (see Bell and Valentine, 1997, p. 125). But at the same time, Starbucks has been a key target of the global justice movement for some very good reasons: because it ostentatiously promotes its participation in fairtrade whilst only using 1% of fairtrade products in practice; because its monopolistic anti-competition tactics aggressively target smaller cafes; and because it is renowned for paying low wages whilst CEO Jim Donald earns more than 2 million dollars a year (Lyons, 2005; Talen, 2003). For me, then, Starbucks manifested enough of the exploitative characteristics of neo-liberalism to consider it a more than legitimate target of hostility and satire.

Alongside this, I liked the 'consumer whore' part of the logo. It was redolent of a wider gendered aesthetic that had sought to reappropriate terms like 'queer' and 'bitch', to knowingly subvert, re-iterate and make them powerful. After all, I had grown up in the kind of campy feminist postmodernist culture where it was practically obligatory to have fridge magnets and postcards featuring reprints of 1940s and 1950s pulp film posters with titles like *Campus Tramp* and *High School Hellcats*. The alternative Starbucks logo was an extension of this riot-grrl-lite aesthetic, particularly given the context of the pierced woman selling it to me. In tandem with this aesthetic I also liked how it satirised a strain of puritan American sexual morality. For as is mildly well known, the Starbucks siren's nipples – clearly apparent on the 'Consumer Whore' version – were removed from the logo in the early 1990s in the States due to the fear of an outcry from the religious right (Fenkl, 2003; Talen, 2003). (Interestingly, Fenkl describes the 'double-tailed siren' as 'a cross between a mermaid and a sheila-na-gig' that can be found as a decorative motif in many European churches and cathedrals; and points out that, historically, 'her suggestive pose, like that of the sheila-na-gig, referred to female sexual mysteries in particular.')

I also liked, in that particular moment of space and time, how my new ownership or connection to that mock logo interpellated *me* as something of a consumer whore. It enabled me to transfer some of the criticism of Starbucks to a criticism of myself as a consumer: for I know that I am too part of this system of consumerism in ways that I do not really want to be, but nonetheless *am*. As Arundhati Roy puts it, this is a world where there is no innocence, in which 'most of us are completely enmeshed in the way the world works. All our hands are dirty' (Roy, 2004, p. 32). Although I did not know it at the time, this kind of reflexivity about action and complicity echoed the designer of *Consumer Whore*'s sentiments. For graphic artist Kieron Dwyer, who drew the mock logo after noticing a sudden proliferation of mounds of unrecycled, Starbucks-branded trash when a branch opened in his neighbourhood, it was not only 'the planetary destruction wrought by this unstoppable consuming machine... its incitement to consume more, and produce more trash' that galvanised him into creating the parody, but also his annoyance with his own Starbucks habit; the design marked his own 'complicity in the capitalist food-chain' (Dwyer, 2001).

Starbucks sued Dwyer for producing the image in 2000, obtaining an injunction forbidding its use for a year. The settlement of the case banned Dwyer from disseminating it through stickers, comic books or t-shirts, or from putting it on or linking it to his website. The image has however been widely circulated on the net, on sites such as *Illegal Art* (the website of an exhibition sponsored by the anti-consumerist magazine *Stay Free!* (2006)) and reproduced in versions such as the one I bought.

Buying this badge was not by any means the most significant act I have made as a form of what Scandinavian social scientists term 'political consumerism' (consumer-oriented activities which intervene in the question and practice of the distribution of power and resources in society) or 'anti-consumerism' (activities explicitly critical of late capitalist consumer culture) (see Micheletti, 2003). It was only a very small moment and there is clearly an irony and a contradiction in the fact that it was *bought* (not least, on a carbon-spewing transatlantic trip). It would be very easy, for example, to argue that my regular organic box of local food, my occasional participation in Trade Justice Movement rallies or my membership of the Co-operative Bank are much more consequential forms of action. But it is significant because it is often in small moments of affective micro-investment that our ongoing identifications and characters can be made. For me, at the time, it clearly ticked a number of boxes that said something to me about who I thought I was (and who I wanted to be). It was a way of foregrounding and negotiating questions of responsibility and of acknowledging that I was part of the consumer process.

But alongside my personal psychosocial investments the badge has a broader significance. For this alternative Starbucks logo clearly makes a connection between femininity, sexuality and a critique of consumerism, and as such indicates some of the complexities that continue to exist around the subject. How consumerism and its critiques are gendered is a question this chapter seeks to revisit and address, as it is a subject which would benefit from being opened up a little more; particularly in the disciplinary area I am part of, cultural studies, where gender and consumption has been the subject of discussion that is both intense and limited. Therefore, in the spirit of what feminist cultural theorists like Donna Haraway (1991) call 'situated knowledge', I start by relating the subject to myself, before reflecting on the status of this subject within the discipline I work in and then moving outwards to identify some broader latent tendencies in contemporary culture – in a series of what are (hopefully) expanding and interrelated circles of analysis. In doing so, this chapter attempts to draw on and highlight residual and emergent academic resources, and offer alternative genealogies, for contemporary subject positions which are simultaneously pro-feminist and anti-consumerist.

Consumerism and/as gendered emancipation

As a student, both undergraduate and postgraduate (and beyond), I was drawn again and again to the burgeoning academic area of consumer culture and gender. It seemed to offer a way of making sense of the world: here was a frame of reference which could take seemingly ordinary acts – being dragged round the shops as a child, wanting and not having, saving up, buying and connecting to wider 'tribes', the pleasure of purchase and display, my mum's job as shop assistant, my own part-time work behind various counters – and offer a way of connecting these apparently mundane parts of experience to larger contexts. In doing so it provided me with a frame for understanding both aspects of myself and my attitudes and some of the bigger social and cultural narratives and attitudes that had shaped me. Grand as it may perhaps sound, the study of consumer culture offered me a means of thinking both about how I was inserted in history, and about a history beyond myself. To know that cultures of consumption were and still in some ways continue to be trivialised or not taken seriously because of what Mica Nava describes as their (often unconscious) association with the feminine (Nava, 1996, pp. 38–76), and from there to be able to think about how they in turn shape social discourse and lived experience, was therefore as important for me as it was for a wide range of other people, both older and younger.

When I began studying consumer culture, I encountered a wide range of cultural theorists who were busy identifying and discussing the process by which mass consumer culture in 'the west' came to be gendered as feminine (for example, Bowlby, 1985, 1993; de Grazia and Furlough, 1996; Nava, 1992, 1996; Radner, 1995). Such work was engaged in unpacking the effects of an historical elision Andreas Huyssen neatly termed through the formulation 'mass culture as woman'. For, as anxiety towards mass consumer culture reached fever pitch in the late nineteenth and early twentieth centuries at the very same time that the lower-middle and working classes were being granted more political power, the masses knocking at the gate, as Huyssen puts it, 'were also women, knocking at the gate of a male-dominated culture' (Huyssen, 1987, p. 47). High art (and in particular modernism) became like production, gendered as serious and male, structurally depending for their status and value on an opposition to feminised mass consumer culture which, to greater or lesser degrees, was derided as frivolous and inferior.

By the time I was a postgraduate in the mid- to late 1990s, the association of consumer goods, spaces and culture with femininity was being thoroughly explored by a wave of feminist critics in sociology, literary theory, cultural studies and cognate disciplines. Subjects from the socio-sexual significance of the department store to the continuing disparagement of 'the commercial' or 'shopping' merely because of its genealogical association with the feminine were inventively and capaciously explored. Like any academic formation, this wave of studies has not been homogenous, containing elements that connected to a number of areas in very different ways. It played a well-known role both in the formation of cultural studies as a discipline and in the imaginative openings-out of feminist cultural analysis.

However, in the spirit of this strain of academic enquiry, it also seems to be a terrain that needs some of its fundamental premises re-examining, particularly given how the cultural landscape has shifted in terms of popular anxieties over corporate power, global warming and transnational sweatshops. Academia to some extent needs to 'catch up' with strands of popular politics (those which began to be most conspicuously reflected, for example, in Naomi Klein's *No Logo* (2000)) by re-engaging with and re-theorising the subject of gender and political consumption and anti-consumerism. To do this, I am arguing, it makes sense to analyse the ways in which 'political consumption' and gender have been, could be – and *are being* – connected, or re-articulated; and to open the debate about gender and consumer culture out further, and wider, than before.

For one significant factor which has often been blocked or bracketed out is the possibility of allowing space for gendered investments in anti-consumerism. For example, take the collection on gender and consumer culture, *The Sex of Things: Gender and Consumption in Historical Perspective*. The remit of the book is described as follows:

> The pre-eminent concern here is not with moral dilemmas... which place oppression or freedom on one side of the equation or other. Instead, the common task... is precisely to capture the immense transformative powers of capitalist-driven consumption as it constantly refashions notions of authentic, essential woman- and mankind.
>
> (de Grazia and Furlough, 1996, p. 8)

Now, *The Sex of Things* is an extremely useful book. But what I find significant here is how it draws the parameters of what it will consider. Why, when talking about consumerism, do we continually need to restrict it to an explication of the cultural fascinations of capitalism's commodities? Why do questions of oppression and freedom need to be thrown out when we consider consumerism's refashionings? Why cannot challenges to late capitalist systems of consumption be *included* in an analysis of gender and consumption?

Of course my answer will be that they can and they should. At the same time, it is also instructive to think *why* and *how* challenges to late capitalist consumer culture are rarely, if ever, present in such analyses of gender and consumer culture. (One exception was Mica Nava's suggestion, in the early 1990s, that feminist sociology and cultural studies think more about consumerism's progressive power through initiatives like green shopping and fair trade (Nava, 1992, pp. 185–199).) In part, as the quotation reveals, the answer lies with the vexed status of 'morality'. In their keenness to avoid prescriptive moralism, theorists have too often evaded or dismissed the question of constructing ethics, challenges and alternatives to contemporary consumerism.

I think there are perhaps two routes we might take at this point. One is that these questions of the 'gendered emancipations' of 'mainstream' consumer culture might be explored in relation to the theoretical status of 'ethics' and 'morality' in more detail. I will come back to this point (via an engagement with Wendy Brown's use of the Nietzschian concept of *ressentiment*) later on. But first we might note that the absence of these questions of political consumption and gender is also structured through a simple lack of connection: because this sphere of debate has not interlinked, to any significant extent, with the issue of the historically gendered involvement of women in radical consumption – in boycotts, protests, strikes and alternative modes of co-operative consumption.

Alternative genealogies

One genealogy we might highlight, then, is the interesting work on the large role women have often played in consumer boycotts, consumer protest and ethical and green consumption. For example, the Swedish political scientist Michele Micheletti notes that women were key in the *Swadeshi* movement in India in the 1940s, which encouraged 'buying Indian' as an integral part of the struggle to abolish colonial rule. Women were also, she points out, pivotal to the meat and milk strikes in the United States around the turn of the twentieth century, as well as to the National Consumer league's 'White Label' campaign, which fought against sweatshop and child labour (Micheletti, 2004, p. 250). Such gendered involvement in historical forms of anti-consumerism is also documented by the cultural historian Lizabeth Cohen in her history of consumer movements in post-war America, *A Consumer's Republic*, who highlights housewives' boycotts against supermarket exploitation (Cohen, 2003, pp. 366–370). To this we might add women's significant role in green consumer protest and the ethical consumer boom from the 1970s (Gabriel and Lang, 1995; see also Hilton, 2003, p. 313 and pp. 337–338).

Micheletti points out that one reason why women have had a conspicuous presence in political consumerism is because it is a form of political participation which exists outside of a traditionally masculine public sphere. (Indeed, it is sometimes exactly for this association that such actions have been derided: Cohen interestingly notes that protests involving women were often dismissed by a scathing press as 'ladycotts' or 'girlcotts' (Cohen, 2003, pp. 369–370).) Boycotts and 'buycotts' (the purposeful buying of 'progressive' brands) are activities which have not generally tended to require much direct contact with mechanisms of state or government, and have therefore been ways in which women have been able to exert power without having to *explicitly* relate to the public sphere from which they were/are historically excluded (Micheletti, 2004, pp. 259–260). In these terms, gendered involvement in political consumerism is a logical consequence of the gendered nature of the public and the private spheres. Micheletti also considers what she calls 'momism': the idea that women are more involved because of their unique role in relation to motherhood. In this formulation, political consumerism is read as an extension of motherhood's 'ethic of care', one that enables women to continue their other social roles, partly as it is a low-risk political involvement and because 'women find less dichotomous involvements attractive' (Micheletti, 2004, pp. 259–260).

What these stories do is remind us that there is another twist in the tale of gender and consumerism *alongside* that of an elision between femininity and mass culture. There is a genealogy in which women and forms of femininity have been very active in attempting to change, to destroy or to create alternatives to systems of consumerism. So besides thinking consumption as a space and a practice in and through which women have become empowered, we can also remember that involvement with forms of anti-consumerism and political consumption, with movements to change the very types and forms of consumer cultures that are possible, and the conditions under which they operate, have also been a set of practices in which women have been directly engaged. This is important as it alters an image of women as historically primarily 'tinkering at the edges of culture' through consumption, even if that role is revalued to emphasise its power; and at the same time there is a sense in which greater knowledge of historical involvement in political consumption might constitute a *cultural resource*, as an area with potential to be drawn on in the present.

Moreover, as Micheletti puts it elsewhere, thinking seriously about political consumerism also 'motivates the crossbreeding of academic disciplines' (Micheletti, 2003, p. 4). In this spirit, we might usefully cross-fertilise these historical points by relating them to contemporary cultural theorisations of gender. If we apply the analyses of gender essentialism developed by poststructuralist theorists such as Judith Butler and Elizabeth Grosz (Butler, 1993, 2004; Grosz, 1995) to Micheletti's schema, we might make some interesting points. For example, while women's gendered involvement in political consumption might be thought of (in Micheletti's terms) as emancipatory because it bridges the historically gendered divisions between public and private spheres, it could also be seen as profoundly limited in its liberatory potential because it continues to rely on an association between women and the domestic realm. In these terms, such actions might be understood as working to *ramify* the message that women can only participate in restricted forms of politics that are based around their identities as consumers. Equally, it is important to make sure that in thinking about the frequently central roles many women have in family units, femininity is not cemented to the family or to the trope of 'Mother Earth' in a reductively essentialising way (particularly given that the figure of the housewife has tended to loom large in the emergent analysis of ethical consumption). Rather, the *potential* of positions such as those Micheletti calls 'momism', or 'a maternal ethics of care', needs to be rigorously qualified as not being an innate or essential set of characteristics.

The queer business of political consumption

Foregrounding the role women have played in political consumerism can therefore act to complicate the relationship between femininity and consumerism. That role quite clearly demonstrates that challenges to or disparagement of 'consumer culture' do not constitute an *a priori* masculine domain. Yet equally, it would clearly be damaging to simply *celebrate* the role of women in challenging the rules and conditions of consumerism. Rather, we might more helpfully view these private sphere-oriented or housewife-oriented traditions as specific and performed roles: as roles that can be a resource for political consumerism, but which simultaneously need to be opened out so that they can be transgressed, extended and joined by other characteristics. This is both so that it is not only 'women' who are exclusively associated with consumerism, and so that the field of possibilities in relation to political or radical modes of consumption does not get constrained to a series of extremely narrowly defined gendered roles.

My reading draws on debates around radicalism and transgression in the field of gender and cultural theory. In *Space, Time and Perversion*, for example, Elizabeth Grosz makes a strong case for the importance of not pigeonholing or reifying a particular gendered or sexualised position as inherently radical:

> Here I explore the possibilities of thinking sexuality, sexual desire, otherwise, in terms other than though also related to those provided by the currently privileged discourses of sexuality – psychoanalytic, Foucaultian, feminist, and queer theory. I attempt to question and distance myself from the self-proclaimed transgressive status assumed by many representatives of the more vulgar positions within queer and lesbian theory, that witnesses representative theorists taking their own lives, pleasures, and desires as standard of the radical, while designating others, even lesbians and queers, as inherently straight by virtue of their difference.
>
> (Grosz, 1995, p. 5)

This paradigm is helpful because it highlights the necessity of not assuming that a single gendered 'role' in relation to political consumerism is more 'radical' or 'transgressive' than another. It would, for example, be narrow and destructive to see the performatively queer seller of punk-oriented 'consumer whore' badges as the radical position to occupy in relation to political consumerism and the housewife who protests for lower milk prices as somehow more reactionary or straight because of her heteronormativity and position within the private sphere. A more capacious reading would be able to tease out the complicated specificities constituting such positions. Equally, it follows that a more thoroughly or substantially transgressive position would involve encouraging the widening out or multiplication of different potential forms of gendered participation in political consumerism.

Some of this kind of 'widening out' of potential different forms of gendered participation in political consumerism can be traced through relatively new forms emerging at present. Examples include the burgeoning role of green fashion pages in women's magazines; the emergence of popular gurus like Safia Minney, founder of fairtrade eco-clothing company People Tree; or the inventive Brooklyn-based anti-consumer magazine edited by Carrie McLaren, *Stay Free!*

These 'widenings out' bring with them their own complexities and contradictions, in terms of both gender and cultural economy. One good example is LA-based trading company American Apparel. Marketing itself as 'sweatshop-free', as its workers get the minimum wage, it breaks with ethical consumption conventions in some ways by using female sexuality to present itself as 'cool' rather than 'worthy', seeking to attract consumers not on the basis of patronage but through style and a sense of being utilitarian, metropolitan and highly sexualised. Its promotional images attempt to 'undo' hyperglossy, superfeminine, airbrushed images, by showing us (supposedly) sexually emancipated women with pores and slight cellulite through Nan Goldin-style amateur realism, and by lightly flirting with a femme/butch aesthetic (its clothes are, for example, co-advertised with *The L Word*, the prime time US drama series about LA-based lesbians). American Apparel's aesthetics, in other words, disrupt political consumerism as a 'maternal act of care' and in this way extend the repertoire of gendered possibilities in contemporary political consumerism. But at the same time, the company clearly has a problematic relationship to feminism, as they 'redo' some very conventional gender dynamics: through the much-vaunted 'ironic' presentation of women-as-sexualised-objects, which does not manage to fully divest itself of misogyny; through the hierarchical power dynamics of the company (which remains staunchly anti-union); and through a CEO who often talks in interviews about his sexual relations with his female employees (see Littler and Moor, 2008). Both opening out and reinscribing what Butler terms 'the heterosexual matrix', its aesthetics and practices simultaneously exhibit what Angela McRobbie has termed, in a broader context, the 'doing and undoing of progressive gender relations' in postfeminism (McRobbie, 2008).

The other side: Hyperconsumerism as *ressentiment*

Thus the gendered history of political consumerism has the potential to provide a different perspective on the relationship between gender, power and consumption: it constitutes a kind of micro 'hidden history' itself, and is a resource to trace to the present. However, it cannot be simplistically celebrated, for whilst gendered challenges to late capitalist consumer cultures have a rich history, they also have a complicated present.

Returning to the two routes I wrote about earlier, I now consider the other side of this divide: the elision between women and the pleasures of consumer culture. I would argue that in order to open out the kinds of gendered emancipations that might be possible in this area, we need to be prepared to be more critical about the more *conservative* (as opposed to emancipatory) aspects of women's involvement in consumption. In many ways, this argument is not new. Many very sophisticated cultural theorists of gender have emphasised how the more 'damaging' aspects of the relationship between femininity and consumerism – personal, cultural and social – are imbricated in complex ways with its various modalities of pleasure and emancipation (for example, Carter, 1997; Winship, 1980). This tradition, I am arguing, might be foregrounded and more closely connected to contemporary debates over excessive consumption and neoliberalism's unequal and unjust trade practices.

One helpful means of augmenting such a project, I think, might be to schematise this tradition of feminine overconsumption in relation to Nietzsche's concept of *ressentiment*. This concept speaks of the 'slave morality' through which a sense of inferiority becomes a prompt to action which is codified as just and moral, and in which righteousness and inferiority act to ramify each other: 'the moralizing revenge of the powerless' augments 'the triumph of the weak as weak' (Brown, 1995, p. 67). This concept has been deployed within feminist theory to enable a critique of a liberal feminism which structures itself through a perpetual dependency on its victim status (and reactive self-affirmation in relation to this), rather than turning to examine the more complex plays of power as a basis for opening up and developing new ethical positions and gendered identities. Elizabeth Grosz, for example, uses it to discuss the hole that a certain type of feminism can dig for itself, adopting a kind of permanent 'slave morality' which imposes a totalising limit on women's expression as 'it must always remain reactive, ... tied to oppression, based on ressentiment' (Grosz, 1995, p. 15). Or as Wendy Brown puts it, more explicitly:

> [M]uch North Atlantic feminism partakes deeply of both the epistemological spirit and political structure of *ressentiment*... [T]his constitutes a good deal of our nervousness about moving toward an analysis as thoroughly Nietzschian in its wariness about truth as postfoundational political theory must be. Surrendering epistemological foundations means giving up the ground of specifically *moral* claims against domination – especially the avenging of strength through moral critique of it – and moving instead into the domain of the

> sheerly political: 'wars of position' and amoral contests about the just and the good in which truth is already grasped as coterminous with power, as always already power, as the voice of power. In William Connelly's words, overcoming the demand for epistemological foundations does not foreclose ethics but opens up alternative ethical possibilities. Apparently lacking confidence in our capacity to work and prevail in such domains of the political and the ethical, feminism appears extremely hesitant about this move.
>
> (Brown, 1995, p. 45)

As she puts it, a more constructive way to jump in practice and in theory is away from *ressentiment* and into investigating or being aware of the complexities of power, the struggles and wars of position around it. Otherwise, identity politics structured by *ressentiment* in effect simply works to 'reverse without subverting this blaming structure' (p. 70).

We might usefully situate the gendered tradition of 'female consumerism as emancipation' within this matrix. For if we insist on focusing on how women's involvement in consumerism is *emancipatory,* we fall into the same trap of uncritically validating a reactive form of liberation, as a form of accruing power that is negative, that is reactive, that is always tied to oppression, to the historical disenfranchisement from production and from the valorised space of the public sphere. The alternative is to develop a more rigorous and capacious ability to consider in what ways consumerism as used by and as affecting women can be detrimental, regressive and conservative as well as emancipatory. To do this is not a means to wallow in misery, to scapegoat women or to position them as cultural dupes. Rather, it is to point out that whilst its crucial not to blame women, it is short-sighted not to recognise the specific nature of their involvement in consumer inequality and overconsumption, issues which have been brought to the foreground of contemporary cultural debate through environmental anxieties and an informational culture which enables the affective detail of neoliberal sweatshop practices to be made more conspicuous. It means that we need to allow ourselves to be more critical of gendered involvement in overconsumerism and exploitative dimensions of consumption/production without either *deploying* or *being frightened of* some kind of residually gendered discrimination.

To emphasise this point now can facilitate arguments about how consumerism itself might be destructive, in both a specifically gendered and an ecological-political sense. It is one means of working towards an articulation between anti-consumerist perspectives *and* the sophisticated conceptual tools we have at our disposal in order to more capaciously recognise the conservative nature of some gendered consumption in the attempt to push beyond it. We might engage in this form of analysis, in other words, and think about the perhaps unpleasant subject of the more conservative forms of gendered consumption, with a slightly different imaginary horizon: one in which the problems of overconsumption and consumerism *coexist*, and *are imbricated with*, problems of gender and inequality.

For example, what we might now term 'gendered overconsumption' can be regarded as possessing a genealogy in its own right, one we might trace all the way up to contemporary modalities of femininity which are encouraged to overspend. There is a very strong feminist tradition that can be re-engaged with and re-emphasised in which women's involvement with consumer culture can be read as profoundly regressive as well as emancipatory. Joan B. Landes, for instance, has shown in her study *Women and the Public Sphere in the Age of the French Revolution* (1988) that middle-class women's increasing interpellation as consumers in the eighteenth century was to a large degree compensation for their being pushed outside the newly established public sphere. Jennifer Jones' recent work on fashion in eighteenth-century France illustrates how consumers were encouraged to accrue more identities and styles both as compensation for this exclusion and as a constituent part of the engine of capitalist growth (Jones, 2004). This kind of analysis, which is suggestive of how the rise of possessive individualism was gendered, relates to Janice Winship's work on how women's fashion magazines in mid-twentieth-century Britain encouraged an ideology of individualism, with the double-edged forms of social and gendered power that this brought. Christine L. Williams' research in the retail sector in the 1990s/2000s includes a discussion of how staff working on the shop floor overwhelmingly considered the most problematic customers to be a strand of middle and upper-middle class white women, whose sense of entitlement and financial and cultural empowerment as consumers frequently made them more likely to treat workers with disdain, as if they were servants (Williams, 2006, p. 109).

From another angle, we might consider how recent work on neoliberal femininities (for example, Hennessey, 2000) has explored some of the complicated pressures as well as pleasures brought through cultures of purchase for a generation of teenagers which is allegedly more consumer-active yet more unhappy than ever before (see Mayo, 2005). These questions are interestingly explored in Angela McRobbie's recent intervention in *Cultural Studies*, which, in discussing new forms of femininity, draws attention to the 'forms of compulsory consumption' which the TV series *Sex in the City* and other programmes have 'so fully endorsed'. As McRobbie points out, some feminist commentators, by seeking to unproblematically 'celebrate' the series as 'fans' and by avoiding the use of theory, have been complicit in this process (she refers in particular to the collection of essays edited by Akass and McCabe, 2003). The programme is paradigmatic of a mode of overconsuming femininity in which emancipation is supposedly delivered to a large extent through hyperconsumption. This is not the only factor we might discuss about *Sex in the City*, but neither is it a factor to be airbrushed out or swept under the carpet – on the contrary, it needs to be more fully integrated into this and similar analyses of gendered identities and their representation.

Equally, such a gendered analysis of consumerism from a more politically aware, anti-consumerist perspective means investigating boys and their toys, or the articulations between 'hyperconsumerism' and masculinities. Such enquiries might draw on the considerable work in the humanities and social sciences which has been concerned with consumerism and masculinity (see, for example, Breward, 1999), or on contemporary resources like the National Consumer Council's report, *Shopping Generation*, which highlights how male teens in the United Kingdom have a pronounced investment in shopping but tend to code their purchasing activities as being more 'essential' and meeting more 'needs' than feminine consumption (Mayo, 2005, pp. 11–12). The question of how forms of masculinities are encouraged to hyperspend, while that spending often remains coded as an 'invisible' or more 'necessary' form of consumption, could be more capaciously related to the context of the global availability or lack of environmental resources. In addition, there is a whole category of consumer goods that are complex and ambiguous in their gendering, particularly the ever-growing array of electronic and digital goods – with their attendant mountains of waste (see Parks, 2007).

Concluding reflections

We have seen that consumption in the West, as a host of commentators have observed, has historically been overwhelmingly gendered as feminine. Interestingly, however, women's prominence in political consumption and anti-consumerist movements and campaigns around alternative, radical and ethical consumption has received less serious attention, despite gestures in this direction. The elision of such a hidden micro-history is partly, I have suggested in this chapter, because it disturbs more simplistic paradigms of consumer culture as a site of feminist liberation which are structured around *ressentiment*. I have argued that we need to become more aware of how gender can contribute to *reproducing* and *extending* both highly conservative forms of consumerism and various forms of political consumption, radical consumption or anti-consumption.

It might be asked, why bother to look at femininities or gender at all as a site for this problematic, and risk re-inscribing women's role as scapegoat for the evils of consumption? Why not look at the issue from more oblique angles – such as the way other goods and services are used (like oil)? Well, yes, we should do those things too. But if we are *ever* going to say anything about the relationship between femininities and consumer culture, then it seems to me that we also need to be ready, first to acknowledge how consumerism has a stronger and more complex internal history of radical consumption and revolt than is often recognised, and second to acknowledge the forms of complicity that have existed and can exist between women and conservative consumerism, calibrated as they are along various specific axes of class and 'race'. If we are ready to discuss these factors, then we will be able to explore how we might open up more subject positions for both gender and political consumption, and might extend the possibilities of both.

Acknowledgements

Thanks to Sara Hackenberg, the editors of this volume, and participants at the 2006 *Crossroads in Cultural Studies* and *Countering Consumerism* conferences for their feedback and support.

References

Akass, Kim and Janet McCabe (eds) (2003) *Reading Sex in the City* London: I.B. Tauris.
Bell, David and Gill Valentine (eds) (1997) *Consuming Geographies* London: Routledge.
Bowlby, Rachel (1985) *Just Looking* London: Methuen.
Bowlby, Rachel (1993) *Shopping with Freud* London: Routledge.
Breward, Christopher (1999) *The Hidden Consumer: Masculinities, Fashion, and City Life 1860–1914* Manchester: Manchester University Press.
Brown, Wendy (1995) *States of Injury* New Jersey: Princeton.
Butler, Judith (1993) *Bodies that Matter* London: Routledge.
Butler, Judith (2004) *Undoing Gender* London: Routledge.

Carter, Erica (1997) *How German is She? Postwar West German Construction and the Consuming Woman* Ann Arbor: University of Michigan Press.
Cohen, Lizabeth (2003) *A Consumers' Republic: The Politics of Mass Consumption in Postwar America* New York: Vintage.
de Grazia, Victoria and Ellen Furlough (eds) (1996) *The Sex of Things: Gender and Consumption in Historical Perspective* Berkeley: University of California Press.
Dwyer, Kieron (2001) 'Sued by the Siren, Part 1: Confessions of a frap addict' http://www.tmcm.com/pages/mag_content/tm0/sued_pt1.html.
Fenkl, Heinz Insu (summer 2003) 'The Mermaid' in *The Endicott Studio Journal of Mythic Arts* http://www.endicott-studio.com/jMA03Summer/theMermaid.html.
Gabriel, Yiannis and Tim Lang (1995) *The Unmanageable Consumer* London: Sage.
Grosz, E. (1995) *Space, Time and Perversion* London: Routledge.
Haraway, Donna (1991) *Simians, Cyborgs and Women* London: Routledge.
Hennessey, Rosemary (2000) *Profit and Pleasure in Late Capitalism* New York: Routledge.
Hilton, Matthew (2003) *Consumerism: The Search for an Historical Movement* Cambridge: Cambridge University Press.
Huyssen, Andreas (1987) *After the Great Divide: Modernism, Mass Culture and Postmodernism* Bloomington: Indiana University Press.
Jones, Jennifer (2004) *Sexing la Mode: Gender, Fashion and Commercial Culture in Old Regime France* Oxford: Berg.
Klein, Naomi (2000) *No Logo* London: Flamingo.
Landes, Joan (1988) *Women and the Public Sphere in the Age of the French Revolution* Ithaca: Cornell University Press.
Littler, Jo and Liz Moor (2008) 'Fourth worlds and neo-fordism: American apparel and the cultural economy of consumer anxiety' *Cultural Studies* 22, pp. 5–6.
Lyons, James (2005) 'Think seattle, act globally: Speciality coffee, commodity biographies and the promotion of place' *Cultural Studies* 19(1), pp. 14–34.
Mayo, E. (2005) *Shopping Generation* London: National Consumer Council.
McRobbie, Angela (2008) 'Young women and consumer culture: An intervention' *Cultural Studies* 22, pp. 5–6.
Micheletti, Michele (ed.) (2003) *Political Virtue and Shopping* London: Palgrave MacMillan.
Micheletti, Michele (2004) 'Why more women? Issues of gender and political consumerism' in Micheletti *et al.* (eds) *Politics, Products and Markets* Brunswick: Transaction.
Nava, Mica (1992) *Changing Cultures: Feminism, Youth and Consumerism* London: Sage.
Nava, Mica (1996) 'Modernity's disavowal: Women, the city and the department store' in Nava and Alan O'Shea (eds) *Modern Times* London: Routledge.
Parks, Lisa (2007) 'Falling apart: Electronics salvaging and the global media economy' in Charles Acland (ed.) *Residual Media* Minneapolis: University of Minnesota Press.
Radner, Hilary (1995) *Shopping Around: Feminine Culture and the Pursuit of Pleasure* New York: Routledge.
Roy, Arundhati (2004) *The Chequebook and the Cruise Missile* London: Harper.
Stay Free (2006) *Illegal Art* http://www.illegal-art.org/index.html.

Talen, Bill (2003) *What Should I do if Reverend Billy is in My Store?* New York: New Press.

Williams, Christine L. (2006) *Inside Toyland* Berkeley: University of California Press.

Winship, Janice (1980) 'Woman becomes an individual – Femininity and consumption in women's magazines 1954–69' in *Stencilled Paper, CCCS,* University of Birmingham.

10

Growing Sustainable Consumption Communities: The Case of Local Organic Food Networks

Gill Seyfang

Sustainable consumption is rising up the environmental policy menu, as a strategy to achieve more sustainable development which requires widespread changes in behaviour at all levels of society to reduce the environmental impacts of consumption (DEFRA, 2003b). While new international environmental governance institutions are growing upwards from state to global scale to tackle system-wide environmental issues, there is an increasing focus upon smaller-scale governance and citizen action at various sub-national levels, from local government to grassroots community groups and individuals (DEFRA, 2005; HM Government, 2005; Seyfang, 2003). New tools are needed to develop and enact these agendas within communities; this chapter examines one such initiative, namely a local organic food system, and assesses its potential role in promoting sustainable consumption.

There is a growing policy emphasis on how motivated individuals can exercise consumer sovereignty and transform markets through everyday purchasing decisions. However, a sociological analysis of consumption suggests that the scope for individuals and groups to change their behaviour is limited by existing social infrastructure and institutions – systems of provision – which 'lock in' consumers into particular patterns of consumption (Levett *et al.*, 2003; Maniates, 2002; Sanne, 2002). 'Systems of provision' are vertical commodity chains (comprising production, marketing, distribution, retail and consumption in social and cultural context) which link 'a particular pattern of production with a particular pattern of consumption' (Fine and Leopold, 1993, p. 4). Within the 'New Economics' literature, sustainable consumption is understood to require fundamental changes in lifestyles, economic and social systems to seek increases in quality of life rather than in material consumption (Jackson, 2004). It therefore demands a deeper understanding of the systems of provision which mediate consumption patterns, in order fundamentally to transform these elements of social infrastructure (Southerton *et al.*, 2004; Van Vliet *et al.*, 2005).

Local organic food provision has been widely advocated as a practical means of making the desired changes to conventional production and consumption systems (see, for example, Douthwaite, 1996; Jones, 2001; Norberg-Hodge *et al.*, 2000). Previous research has studied the economic and social impacts of relocalised and alternative food networks (DuPuis and Goodman, 2005; Holloway and Kneafsey, 2000; Murdoch *et al.*, 2000; Renting *et al.*, 2003; Winter, 2003), and the environmental implications of local versus imported food, and organic versus conventionally produced food (For example, Pretty, 2001). To date, however, there has been no systematic appraisal of local organic food as a strategy for sustainable consumption, and no suitable evaluation frameworks have yet been developed. Here, I present an empirical evaluation of a local organic food network as a tool for sustainable consumption, aiming to contribute to the debate on environmental governance by discussing the role and potential of local organic food networks in developing new institutions which enable individuals and groups to change their consumption patterns.

I begin by setting out the rationale for the New Economics model of sustainable consumption, and the role within that of local organic food systems. New Economics theory provides qualitative criteria to evaluate the effectiveness of initiatives in achieving sustainable consumption. These are applied to a study of an organic producer cooperative in Norfolk, United Kingdom, and the research findings are presented. Finally, ways forward for community-based sustainable consumption are discussed, with appropriate policy recommendations.

Mainstream and alternative visions of sustainable consumption

Responsibility for environmental decision-making in its widest sense is shifting from central government to new actors and institutions, at a range of scales (Jasanoff and Martello, 2004). Over the last 15 years, 'sustainable consumption' has become a core issue on the international environmental agenda (OECD, 2002; UNCED, 1992). In 2003, the UK Government announced its strategy for sustainable consumption and production, defining this as 'continuous economic and social progress that respects the limits of the Earth's ecosystems, and meets the needs and aspirations of everyone for a better quality of life, now and for future generations' (DEFRA, 2003b, p. 10). In practice, this emphasises decoupling economic growth from environmental degradation, through a range of market-based measures, and calling on informed and motivated citizens to use their consumer sovereignty to transform markets by demanding improved environmental and social aspects of production and product design (ibid.).

This mainstream policy approach to sustainable consumption has been criticised – not least by the government's own Sustainable Development Commission – on the basis of a number of significant factors which critics claim limit its effectiveness and scope (Porritt, 2003). These are that it relies upon market signalling, which in turn is based upon pricing regimes which systematically externalise social and environmental costs and benefits; that it fails to consolidate (in policy) improvements made over time, leaving them vulnerable to changes in consumer attention and concern; that it makes only consumer markets available to transformation, leaving significant consumption from producer industries, and institutional consumption through the public sector, immune to sustainable consumerism; that it neglects the social meanings and context of consumption which compete for influence with environmental motivation; that it affords the right to influence the market solely to those able and willing to participate in that market; that it cannot encompass action to reduce consumption and seek alternative channels of provision such as informal exchange networks by consumers eager to create institutions representative of their values; that it pits individuals against globally powerful corporations in an inequitable struggle; and most significantly, that it fails to see the social infrastructure and institutions which constrain choice to that available within current systems of provision. The critics therefore conclude that the mainstream approach is limited in scope, flawed in design and unjust in its objectives (Burgess *et al.*, 2003; Holdsworth, 2003; Levett *et al.*, 2003; Maniates, 2002; Sanne, 2002; Seyfang, 2004, 2005; Southerton *et al.*, 2004).

If current systems of provision prevent significant changes in consumption patterns, what can be done to overcome this limitation? Alternative systems of provision, with associated social and economic institutions and infrastructure, require a foundation in alternative values, development goals, motivations and definitions of wealth (Leyshon *et al.*, 2003). Advocates of an alternative approach draw out the political economy of consumption, note its wider social meanings and point to collective institutions as the source of potential change (Fine and Leopold, 1993; Maniates, 2002). Such an alternative approach is proposed by a broad body of thought known as the 'New Economics' (Boyle, 1993; Daly and Cobb, 1990; Ekins, 1986; Henderson, 1995). Here, I explore the practical social implications of this normative theory.

A New Economics evaluation framework for sustainable consumption

The New Economics is an environmental philosophical and political movement founded on a belief that economics cannot be divorced from its foundations in environmental and social contexts (Lutz, 1999). It emerged from the environmental movement and built upon the work of green writers such as Schumacher and Robertson to develop a body of theory about how a 'green' economics concerned with justice and social well-being could be envisioned and practised. The UK's New Economics Foundation (a self-styled 'think-and-do-tank') was founded in 1986 to promote these ideas in research and policy (Ekins, 1986), and is now the leading think tank concerned with developing practical knowledge and skills in this area, and feeding these ideas into policy. At the same time, theorists such as Jackson, Ekins, Douthwaite and O'Riordan are pursuing these ideas within the academic world, for instance by developing new measures of well-being, seeking to understand consumer motivations in social context, and debating how an 'alternative' sustainable economy and society might operate. Nevertheless, despite a growing number of practical applications of this model, there is a paucity of robust empirical research to test the New Economics approach, and there has been no systematic means of evaluating activities to assess their contribution to sustainable consumption. The new qualitative evaluation framework proposed here is designed to incorporate the key elements of the New Economics vision of sustainable consumption. There are five key points.

The first of these is *localisation*. New Economics stresses the benefits of social and economic decentralisation and local self-reliance in order to protect local environments and economies from external shocks and the negative impacts of globalisation (Jacobs, 1984; Schumacher, 1993). It proposes an 'evolution from today's international economy to an ecologically sustainable, decentralizing, multi-level one-world economic system' (Robertson, 1999, p. 6), or what is known today as the 'new localism' (Filkin *et al.*, 2000). Localisation need not imply autarky: rather, products should be produced as close to the place of consumption as is reasonably possible, and meeting needs locally should be given greater prominence in economic development. Building stronger localised economies is therefore a priority, and can occur through increasing the economic multiplier (the number of times money changes hands before leaving an area), which in turn occurs as a by-product of import-substitution or local provisioning (Douthwaite, 1996).

Secondly, sustainable consumption demands an equitable distribution of environmental goods and services, which requires developed countries to *reduce their ecological footprints*. Taking an equity-based understanding of environmental governance and global interdependence, the New Economics draws on 'ecological footprinting' methodology to interpret the impacts of one group of global citizens on others. This defines and visualises environmental injustice in terms of the inequitable distribution of 'ecological space' (the footprint of resources and pollution-absorbing capacity) taken up by individuals, cities and countries (Wackernagel and Rees, 1996). For instance, it is estimated that for the whole world's population to achieve a UK lifestyle would require a total of 3.1 Earths, and that on 16 April in any year, UK citizens have used up their fair share of resources for the calendar year, and begun living off the resources of others (Simms *et al.*, 2006). Redressing this inequitable distribution requires a reduction in material consumption in affluent economies, through recycling, reducing demand, sharing facilities and resources and so on. This would be accompanied by a reorientation of economic development goals away from production and consumption measures (such as Gross Domestic Product) and towards measures of well-being – which is increasingly found to not correlate with material consumption above a certain income level (Jackson, 2004; Layard, 2006).

The third factor is that of *community-building*. This approach calls for a new 'ecological citizenship' of all humanity, a community which expands across borders (as does environmental change) and recognises the political implications of private decisions, and so defines everyday activities of consumption as potentially citizenly work (Dobson, 2003). At the same time, it advocates resilient, inclusive, diverse local communities of place and interest to provide sustainable places to live and work (Barton, 2000). Overcoming social exclusion, nurturing social capital and developing active citizenship within participative communities are key aspects of this (O'Riordan, 2001).

Fourth, and emerging from a basis in sustainable communities, the New Economics approach emphasises the potential for *collective action* to overcome the powerlessness and individualisation of responsibility inherent in the mainstream model (Maniates, 2002). This includes the possibility of acting collectively to influence decisions and deliver services through political decision-making processes, and also addresses questions of institutional consumption – through the public sector, for example (Seyfang and Smith, 2007).

Finally, and following from this last point, perhaps the most important outcome of collective action is the potential to create *new socio-economic institutions* – alternative systems of provision – based upon different conceptions of value. New Economics redefines 'wealth', 'prosperity' and 'progress' in order to construct new social and economic institutions for governance which value the social and environmental aspects of well-being alongside the economic (Jackson, 2004). Since current systems of provision limit the effective choices available to individuals, constructing new social infrastructure according to alternative values allows people to behave as ecological citizens (Seyfang, 2005, 2006).

These indicators provide a number of criteria to evaluate sustainable consumption. Depending on the case to which they are applied, there may be some overlap between them, notably the last three. However, each captures distinct aspects of consumption. In what follows, these criteria are applied to a local organic food cooperative, a system of food provisioning put forward by proponents of the New Economics which is claimed to promote sustainable consumption. Before considering the particular initiative, let us review the New Economics rationale for local organic food.

The rationale for local and organic food networks

Organic production refers to agriculture which does not use artificial chemical fertilisers and pesticides, and in which animals are reared in more natural conditions, without the routine use of drugs, antibiotics and wormers common in intensive livestock farming. The first sustainable consumption rationale for organic food is that its production is more in harmony with the environment and local ecosystems. By working with nature rather than against it, and replenishing the soil with organic material, rather than denuding it and relying upon artificial fertilisers, proponents claim that soil quality and hence food quality will be improved (with benefits for consumer health and food safety), biodiversity will be enhanced, and farmers can produce crops without large-scale industrial chemical inputs, which pollute waterways and degrade the land (Reed, 2001). The area of land within the UK certified and in conversion for organic production has risen dramatically in recent years: from under 100,000 hectares in 1998, this had risen by 2003 to 741,000 hectares (DEFRA, 2003a). However, while this rapid expansion signifies a growing demand for less environmentally damaging food production, Smith and Marsden point out that the sector may be heading towards a 'farm-gate price squeeze' common within conventional agriculture, which will limit future growth and development. Farmers keen to diversify into organic production to secure more sustainable livelihoods in the face of declining incomes within the conventional sector confront an efficient supermarket-driven supply chain which increasingly sources its organic produce from overseas. Currently 65% of organic produce eaten in the United Kingdom is imported, and 82% is sold through supermarkets (Soil Association, 2002). A key challenge for small organic producers is therefore to create new distribution channels to bypass the supermarket supply chain, and organise in such a way as to wield sufficient power in the marketplace.

One way to achieve this is through the promotion of specifically *local* organic food, to nurture a new sense of connection with the land, through a concern for the authenticity and provenance of the food we eat – which is a social as much as a technological innovation (Smith, 2006). This movement towards the (re)localisation or shortening of food supply chains explicitly challenges the industrial farming and global food transport model embodied in conventional food consumption channelled through supermarkets (Reed, 2001). The explosion of farmers' markets, direct marketing, regional marketing and similar initiatives has supported this turn towards 'quality', 'authentic', 'relocalised' local food (Holloway and Kneafsey, 2000; Murdoch *et al.*, 2000; Ricketts Hein *et al.*, 2006). The main environmental rationale for localising food supply chains is to reduce the social and environmental impacts of 'food miles' (the distance food travels from production to consumption). Much transportation of food around the globe, and its attendant carbon dioxide emissions, is only economically rational because environmental and social externalities are excluded from fuel pricing (Jones, 2001). It results in the sale of vegetables and fruit from across the globe, undercutting or replacing seasonal produce in the United Kingdom. Pretty (2001) calculates the cost of environmental subsidies to the food industry, and compares the 'real cost' of local organic food with that of globally imported conventionally produced food. He finds that environmental externalities add 3.0% to the cost of local-organic food, and 16.3% to the cost of conventional-global food. A report commissioned by the UK government into the usefulness of the 'food miles' concept finds that the direct environmental, social and economic costs of food transport are over £9 billion each year, over £5 billion being attributable to traffic congestion (Smith *et al.*, 2005).

In addition, there are social and economic rationales for relocalised food supply chains within a framework of sustainable consumption. Whereas the globalised food system divorces economic transactions from social and environmental contexts, the New Economics favours 'socially embedded' economies of place, developing connections between consumers and growers, and strengthening local economies and markets against disruptive external forces (Norberg-Hodge *et al.*, 2000). Indeed, rather than being eroded by globalisation, these diverse embedded food networks are now flourishing as a rational alternative to the logic of the global food economy (Whatmore and Thorne, 1997). They are making a significant contribution to rural development, mitigating the crisis of conventional intensive agriculture, and mobilising new forms of association which might resist the price-squeeze mentioned above (Renting *et al.*, 2003). This is demonstrated in a study of food supply chains in Norfolk which found that the motivations for many growers to sell locally included 'taking more control of their market and [becoming] less dependent on large customers and open to the risk of sudden loss of business' (Saltmarsh, 2004, Chapter 3). Many of these growers faced constant insecurity over sales, and turning towards the local market was a means of stabilising incomes and self-protection. As well as protecting farmers, localisation builds up the local economy by increasing the local multiplier (Ward and Lewis, 2002).

Localism is not uncritically embraced within the New Economics. Localisation can be a reactionary and defensive stance against a perceived threat from globalisation and different 'others' (Hinrichs, 2003; Winter, 2003), and the local can be a site of inequality and hegemonic domination, far from conducive to the environmental and social sustainability often attributed to processes of localisation by activists. Thompson and Arsel (2004) describe such uncritically pro-localisation consumers as 'oppositional localists', who see only positive characteristics in small-scale local organisations and businesses, and completely reject globalised business. Their research indicates the need to be objective about consumers' motivations and underlying values. Moreover, localism raises questions of 'sustainability for who?', as the nascent desire for locally produced food in developed countries inevitably affects the economic and social destinies of food-exporting developing countries. New Economists argue for a globalised network of local activism which addresses the economic and social needs of developing countries reliant upon food exports, and prioritises fair trade for products which cannot be produced locally, while simultaneously lobbying for international trade justice. A reflexive localism offers ecological citizens the opportunity to forge both local and global alliances with progressive actors at the local level and consciously avoid the negative aspects of defensive localism (DuPuis and Goodman, 2005).

I now turn to an empirical case study of a local organic food provisioning system, Eostre Organics. Using the five criteria for sustainable consumption outlined above, this assesses the activities of an organic producer cooperative, and the motivations of its organisers and customers, to explore how far it is an effective vehicle for sustainable consumption. The mixed-method study involved a site visit to the headquarters and main box-packing area; several site visits to the market stall; semi-structured interviews with the founder, the marketing and development officer, two growers and the market stallholder; document analysis of Eostre's website and newsletters; and self-completion questionnaires for consumers. These surveys, asking customers about their attitudes to organic and local food, were sent to all 252 customers of three of the veggie-box schemes supplied by Eostre. Seventy-nine were returned, representing a response rate of 31%. In addition, all customers of the Norwich market stall were invited to take a survey; 110 did so, and of these 65 were returned (59% response rate). Market stall staff reported that while not every customer took a survey during the two-week period when they were available, most of their regular customers did. Overall, the survey achieved a 39% response rate. It sought both quantitative and qualitative information on the consumption patterns, values and motivations of Eostre's customers, aiming to elicit a wide range of views and to allow consumers to express in their own terms how they understand and respond to food consumption issues. As well as noting quantitative findings of the survey, I include a range of quotations from customers' responses.

Evaluating Eostre Organics, a local food cooperative

Eostre Organics is an organic producer cooperative based in Norfolk, in the East of England, established in 2003 with development funding from DEFRA's Rural Enterprise Scheme. Eostre comprises nine local organic growers – some with very small holdings – and a producer cooperative in Padua, Italy with over 50 members of its own. These farms produce a wide range of seasonal fruit and vegetables, and local supplies are supplemented (but not replaced) by imports from their Italian partners and other co-operative and fair trade producers. Produce is sold through a full-time market stall, weekly subscription box schemes, shops, and farmers' markets, and local schools and a hospital are supplied.

Eostre's charter states:

> Eostre believes that a fair, ecological and co-operative food system is vital for the future of farming, the environment and a healthy society. Direct, open relationships between producers and consumers build bridges between communities in towns, rural areas and other countries, *creating a global network of communities, not a globalised food system of isolated individuals.*
>
> (Eostre Organics, 2004; emphasis added)

Eostre's organisers are clearly motivated by ecological and social objectives, but how successful are they at achieving them? Taking each of the five criteria for sustainable consumption in turn, we now examine the practices and perceptions of producers and consumers in this alternative food system, to assess its effectiveness at achieving sustainable consumption.

Localisation

The principal aim of Eostre was to support the livelihoods of local organic producers within the region by enabling them to serve local markets. This aim has been achieved so far: Eostre saw a 70% increase in sales during the first year of operation, and has expanded its range of retail outlets. Indeed, an index of food relocalisation developed by Ricketts, Hein and others ranks Norfolk ninth among the 61 counties of England and Wales. Consumers also value local producers highly: 84% of the survey respondents said they chose Eostre because of a commitment to supporting local farmers. One consumer said, 'I value the fact that some of it is grown in Norfolk by small businesses whose owner and workers obviously care about the land, their customers and their social surroundings.' Another stated, 'I would like to see a return to seasonal fruit and veg, which we can only hope for if we support the smaller local farms.' Keeping money circulating in the local economy by patronising locally owned businesses was a motivation for 65% of consumers who responded: one said, 'We like to support local growers and local industry.' The theme of self-reliance was also prominent: 'I like the idea of England being more self-sufficient and using our own good land to feed us all simply', said one respondent, and 36% wanted to preserve local traditions and heritage through supporting Eostre.

'Food miles' was a concept important to Eostre's customers when thinking about localisation. Eostre's marketing manager explains,

'People are becoming very eco-aware, and one of the biggest issues in any ecological awareness has got to be food miles.' This is supported by the survey which found that 84% of respondents specifically aimed to reduce food miles through buying from Eostre. Typical explanations included, 'If good, tasty food is available locally, it seems pointless to buy potentially inferior goods from a supermarket which have often been imported from across the globe'; 'It cuts out the environmentally-destructive chain of transport from one end of the world to another'; and 'It supports the local economy, reduces food miles, and enhances the local countryside.' However, at present consumers sometimes face a trade-off between local and organic attributes of their food, and must choose, according to their priorities, between conventionally produced local food and imported organic produce. One customer stated, 'I don't believe [imported] organic is worth the food miles.' Eostre currently supplements its range with imported organic produce, where gaps exist, but an increase in local production capacity would help to fill many of those gaps.

Reducing ecological footprints

A commitment to sustainable farming and food is evident in Eostre's mission statement, and this is forcefully supported by their customers. Of the customers who responded to the survey, 94% stated that they bought from Eostre because they believed local and organic food was better for the environment. One respondent said, '[Buying local organic food] is important because we believe in sustainability regarding our environment, and we are committed to reducing our "eco-footprint" in any areas we can.' Another stated, 'I feel I owe it to the Earth.' A third explained, 'I am very concerned about the effects of pesticides and pollution on us and the environment.' Another was motivated by the fact that 'organic farming is better for wildlife'. Respondents' statements suggest that the environmental factors being considered are farm-related (pesticide and fertiliser use), transport-related (food miles) and packaging-related (85% of respondents chose Eostre in order to reduce unnecessary food packaging). Another customer explained, 'To me, it represents a more harmonious ecological balance between that which we produce, consume and waste.'

Community-building

As well as strengthening the local economy and reducing environmental impacts, Eostre is a community-building initiative. Links are built up between farmers and consumers, and consumers gain a sense of connection to the land through the personal relationships which develop. One Eostre customer appreciatively noted 'the sense of communal participation, starting from the feeling that we all know – or potentially know – each other, and continuing on through wider issues, both social and environmental.' Another stated, 'I feel that "connectedness" is important', while another reported that they liked Eostre because 'it's a cooperative; they are like-minded people'. These personal connections are developed through face-to-face contact on the market stalls or with box-deliverers, and through newsletters which share stories, recipes and news about the farms and invite customers on educational farm visits. Three quarters (76%) of those customers who completed the survey reported that they were motivated to purchase from Eostre because they liked to know where their food had come from, and a quarter (25%) specifically liked the face-to-face contact with growers. This sense of community is echoed by another respondent who favours local organic food because 'purchasing it links me with a part of the community which operates in a far healthier and more ethical way than the wider economic community'. Another felt that organic food 'helps bring back small community living instead of alienated individuals feeling unconnected'.

Local organic food networks build community and shared vision, and the Eostre market stall in Norwich is a good example of how this works: it is a convenient city-centre meeting point and source of information, open to everyone. The stall is decorated with leaflets and posters advertising a range of sustainable food and other environmental initiatives – for example anti-GM meetings, Green Party posters, alternative healthcare practices and wildlife conservation campaigns. This accurately reflects customers' interests: 60% of respondents identified the Greens as the political party which best represented their views. But how socially inclusive is this community? Organic food is often dismissed as the preserve of an élite, on grounds of price, and claimed to be inaccessible to lower-income groups (Guthman, 2003). In fact many of Eostre's customers are from lower-income brackets. Fourteen per cent of survey respondents had a gross weekly household income of less than £150 (£7,800 a year), compared to 15% of the local population. Higher-income households were under-represented: only 17% of Eostre customers had household incomes of over £750 a week (£39,000 a year), compared to 23% of the local population (ONS, 2003). Only 8% of customers felt that eating organic food reflected 'taste and refinement', suggesting that for most of them, organic is not 'posh nosh'. With such a high proportion of low-income customers, Eostre is achieving its aim of making fresh organic produce available to all social groups.

Collective action

There are two ways in which Eostre is an expression of collective action for sustainable consumption. The first is through its cooperative structure. Many of the farmers in the cooperative had previously sold organic produce to supermarkets, and had suffered from a drop in sales and prices during the recession in the early 1990s, as well as having a negative experience of dependency upon a single, distant buyer. This led some growers to seek greater control over their businesses by moving into direct marketing, and an informal intertrading arrangement developed between a handful of small local organic growers, which formed the core of the cooperative. Eostre aims to provide sustainable, stable livelihoods to its member growers, as a grassroots response to economic recession and vulnerability caused by a global food market – a local adaptation to globalisation in the food sector. By organising collectively, Eostre's members achieve the scale necessary to access markets which small growers cannot satisfy individually: for example, Eostre can supply market stalls all year round. The cooperative values were supported by customers. Seventy per cent of respondents said they chose to buy from Eostre in order to support a cooperative, and one stated, 'I like that local organic farmers work together rather than competing against each other for profit.'

The second collective action impact is through Eostre's inroads into public sector catering through small-scale initiatives, such as providing food for a primary school kitchen and supplying the local hospital's visitors' canteen. These were important first steps, against the ingrained habits and beliefs of public sector catering managers and institutional barriers such as the lack of a kitchen to feed patients in hospitals (cook-chill food being the norm). However, the changing public agenda on school meals resulting from Jamie Oliver's 'School Dinners' TV programme has thrust local organic food provision into the limelight. Eostre and parent NGO East Anglia Food Links (EAFL) have been identified as pioneers with important lessons to share. Currently, heads of catering from seven of the ten East of England Local Education Authorities have agreed to work together with EAFL on a programme to increase the use of sustainable and local food in their school meals (EAFL, 2005).

New institutions

The successes Eostre has achieved in the previous four categories add up to more than the sum of their parts: together they comprise the seeds of a new system of food provision, based upon cooperative and sustainability values (such as fair trade), and bypassing supermarkets to create new infrastructures of provision through direct marketing. Furthermore, their consumers actively support this activity. Many said they were pleased to avoid supermarket systems of provision: 'I think that supermarkets are distancing people from the origins of food and harming local economies; I try to use supermarkets as little as possible'; '[Eostre is] an alternative to a system which rips off producers [and] the planet'; 'I believe in a local food economy.' As one said, 'I don't want supermarket world domination, extra food miles, packaging, and middle people making money!'

The consumer values expressed in these new institutions are quite different to those in mainstream systems of provision. Customers appear to be internalising calculations about social and environmental costs of conventional food production and transport, and responding to more sophisticated, inclusive price incentives than those of the marketplace. One stated, 'I like to pay the "real cost" for my food', and another commented: 'While [Eostre's produce is] not always as cheap as supermarket produce, I am more comfortable knowing that a greater proportion of my money goes to the primary producers.' Another difference is that customers embrace seasonality and accept certain foodstuffs will not be available for several months each year. Subscribers to the box schemes do not even have a free choice over what food they will receive. They are given a box of mixed seasonal fruit and vegetables each week: one likened the inherent surprises to 'having a Christmas present every week! I never know what the box will contain, it's a challenge to my cooking skills!' Others echoed this pleasure in adapting to seasonal availability. A temporary lack of variety might be seen as a major failing in mainstream systems (the vision of empty supermarket shelves inducing panic!), but here it is welcomed as showing connection with the seasons and locality. One customer remarked, 'I reject the ethos of the supermarket that all products should be available all year round. I enjoy the seasonal appearance of purple sprouting broccoli, asparagus, and so on.' Many comments referred to creating new sustainable food systems, confirming that Eostre is beginning to create new provisioning institutions.

The consumers appeared to back up these principles with action. The average weekly household expenditure of survey respondents on all food and drink was £71; of this, over half (£37 or 52%) was spent on local, organic or fairly traded products (from all sources, not just Eostre). Three quarters (75%) of respondents reported that they bought some of this produce from supermarkets, and consumers cited a variety of advantages and disadvantages of supermarket provision over Eostre. They enjoyed the convenience and availability of supermarket organics and local food, but nevertheless retained a general antipathy towards the mainstream supermarket system *per se*. This indicates that for most people, food provisioning is not an all-or-nothing choice, but involves a plurality of approaches and systems, reflecting perhaps the trade-offs between affordability, accessibility and ethics.

Conclusions: Growing sustainable consumption communities

Community action for sustainable development is a growing area of political and practical interest, but empirical research is needed to systematically evaluate its effectiveness and understand the processes through which it takes place. Here, a new evaluation framework for sustainable consumption, based on New Economics theories, has been developed and applied to a case study of a local organic food cooperative. We can conclude that the initiative was successful in its aims of enabling and promoting sustainable consumption, as measured by the key indicators of localisation, reducing ecological footprints, building communities, acting collectively, and building new institutions. Given a conducive policy framework and appropriate development funding, local organic food initiatives could deliver substantial everyday changes in behaviour and environmental impact, while building new and more sustainable social infrastructure and systems of food provisioning. Eostre's customers strongly supported these values and goals. These research findings indicate that local organic food networks like the one discussed enable consumers to enact their non-mainstream values of ecological citzienship, and to join forces with like-minded people in building an alternative to globalised, mainstream food supply chains.

Acknowledgements

This chapter originally appeared (in a slightly different form) as a paper in the *International Journal of Sociology and Social Policy*, Volume 27 (3/4), pp. 120–134. It is republished here with permission from Emerald Group Publishing Limited.

The research was funded by the Economic and Social Research Council as part of CSERGE's Programme on Environmental Decision-Making. Thanks to Beth Brockett for research assistance.

References

Barton, H. (ed.) (2000) *Sustainable Communities: The Potential of Eco-Neighbourhoods* London: Earthscan.

Boyle, D. (1993) *What is New Economics?* London: New Economics Foundation.

Burgess, J., Bedford, T., Hobson, K., Davies, G. and Harrison, C. (2003) '(Un)sustainable Consumption', in F. Berkhout, M. Leach and I. Scoones (eds) *Negotiating Environmental Change: New Perspectives from Social Science* Cheltenham: Edward Elgar, pp. 261–291.

Daly, H. and Cobb, J. (1990) *For the Common Good* London: Greenprint Press.

DEFRA (2003a) *Agriculture in the United Kingdom 2003* London: Stationery Office.

DEFRA (2003b) *Changing Patterns: UK Government Framework for Sustainable Consumption and Production* London: DEFRA.

DEFRA (2005) *Delivering Sustainable Development at Community Level* www.sustainable-development.gov.uk/delivery/global-local/community.htm accessed 24 October 2005.

Dobson, A. (2003) *Citizenship and the Environment* Oxford: Oxford University Press.

Douthwaite, R. (1992) *The Growth Illusion* Bideford, UK: Green Books.

Douthwaite, R. (1996) *Short Circuit: Strengthening Local Economies for Security in an Unstable World* Totnes, UK: Green Books.

DuPuis, M. and Goodman, D. (2005) 'Should we go "home" to eat?: Toward a reflexive politics of localism' *Journal of Rural Studies* 21(3), pp. 359–371.

EAFL (East Anglia Food Link) (2005) *Local Education Authorities Collaborating on Local Food* <http://www.eafl.org.uk/default.asp?topic=SpiceSeven> accessed 10/5/06.

Ekins, P. (ed.) (1986) *The Living Economy: A New Economics in the Making* London: Routledge.

Eostre Organics (2004) *The Eostre Organics Charter* <http://www.eostreorganics.co.uk/charter.htm>accessed 30/3/04, copy on file.

Filkin, G., Stoker, G., Wilkinson, G. and Williams, J. (2000) *Towards a New Localism* London: New Local Government Network.

Fine, B. and Leopold, E. (1993) *The World of Consumption* London: Routledge.

Guthman J. (2003) 'Fast Food / Organic food: Reflexive tastes and the making of "yuppie chow"' *Social and Cultural Geography* 4(1), pp. 45–58.

Henderson, H. (1995) *Paradigms in Progress: Life Beyond Economics* San Francisco: Berrett-Koehler Publishers.

Hinrichs, C. C. (2003) 'The practice and politics of food system localization' *Journal of Rural Studies* 19, pp. 33–45.

HM Government (2005) *Securing the Future: Delivering UK Sustainable Development Strategy* Norwich: The Stationery Office.

Holdsworth, M. (2003) *Green Choice: What Choice?* London: National Consumer Council.

Holloway, L. and Kneafsey, M. (2000) 'Reading the space of the farmer's market: A case study from the United Kingdom' *Sociologica Ruralis* 40, pp. 285–299.

Jackson, T. (2004) *Chasing Progress: Beyond Measuring Economic Growth* London: New Economics Foundation.

Jacobs, J. (1984) *Cities and the Wealth of Nations: Principles of Economic Life* London: Random House.

Jasanoff, S. and Martello, M. (2004) *Earthly Politics: Local and Global in Environmental Governance* Cambridge, Ma: MIT Press.

Jones, A. (2001) *Eating Oil: Food Supply in a Changing Climate* London: Sustain, and Newbury: Elm Farm Research Centre.

Layard, R. (2006) *Happiness: Lessons from a New Science* London: Penguin.

Levett, R., with Christie, I., Jacobs, M. and Therivel, R. (2003) *A Better Choice of Choice: Quality of Life, Consumption and Economic Growth* London: Fabian Society.

Leyshon, A., Lee, R. and Williams, C. (eds) (2003) *Alternative Economic Spaces* London: Sage.

Lutz, M. (1999) *Economics for the Common Good: Two Centuries of Social Economic Thought in the Humanistic Tradition* London: Routledge.

Maniates, M. (2002) 'Individualization: Plant a tree, buy a bike, save the world?', in T. Princen, M. Maniates and K. Konca (eds) *Confronting Consumption* London: MIT Press, pp. 43–66.

Murdoch, J., Marsden, T. and Banks, J. (2000) 'Quality, nature and embeddedness' *Economic Geography* 76(2), pp. 107–125.

Norberg-Hodge, H., Merrifield, T. and Gorelick, S. (2000) *Bringing the Food Economy Home: The Social, Ecological and Economic Benefits of Local Food* Dartington: ISEC.

O'Riordan, T. (2001) *Globalism, Localism and Identity: Fresh Perspectives on the Transition to Sustainability* London: Earthscan.

OECD (2002) *Towards Sustainable Consumption: An Economic Conceptual Framework*, ENV/EPOC/WPNEP(2001)12/FINAL, Paris: OECD.

ONS (Office of National Statistics) (2003) *Regional Trends 38* London: The Stationery Office.

Porritt, J. (2003) *Redefining Prosperity: Resource Productivity, Economic Growth and Sustainable Development* London: Sustainable Development Commission.

Pretty, J. (2001) *Some Benefits and Drawbacks of Local Food Systems*, briefing note for TVU/Sustain AgriFood Network, 2 November, 2001.

Reed, M. (2001) 'Fight The Future! How the contemporary campaigns of the UK organic movement have arisen from their composting past' *Sociologica Ruralis* 41(1), pp. 131–145.

Renting, H., Marsden, T. and Banks, J. (2003) 'Understanding alternative food networks: Exploring the role of short food supply chains in rural development' *Environment and Planning A* 35, pp. 393–411.

Ricketts Hein, J., Ilbery, B. and Kneafsey, M. (2006) 'Distribution of local food activity in England and Wales: An index of food relocalisation' *Regional Studies* 40(3), pp. 289–301.

Robertson, J. (1999) *The New Economics of Sustainable Development: A Briefing for Policymakers* London: Kogan Page.

Saltmarsh, N. (2004) *Mapping the Food Supply Chain in the Broads and Rivers Area* Watton, UK: East Anglia Food Link.

Sanne, C. (2002) 'Willing consumers – or locked-in? Policies for a sustainable consumption' *Ecological Economics* 42, pp. 273–287.
Schumacher, E. F. (1993) *Small is Beautiful: A Study of Economics as if People Mattered* London: Vintage.
Seyfang, G. (2003) 'Environmental mega-conferences: From Stockholm to Johannesburg and Beyond' *Global Environmental Change* 13(3), pp. 223–228.
Seyfang, G. (2004) 'Consuming values and contested cultures: A critical analysis of the UK Strategy for sustainable consumption and production' *Review of Social Economy* 62(3), pp. 323–338.
Seyfang, G. (2005) 'Shopping for sustainability: Can sustainable consumption promote ecological citizenship?' *Environmental Politics* 14(2), pp. 290–306.
Seyfang, G. (2006) 'Ecological citizenship and sustainable consumption: Examining local food networks' *Journal of Rural Studies* 40(7), pp. 781–791.
Seyfang, G. and Smith, A. (2007) 'Grassroots innovations for sustainable development: Towards a new research and policy agenda' *Environmental Politics* 16(4), pp. 584–603.
Simms, A., Moran, D. and Chowla, P. (2006) *The UK Interdependence Report* London: New Economics Foundation.
Smith, A. (2006) 'Green niches in sustainable development: The case of organic food in the UK' *Environment and Planning C* 24, pp. 439–458.
Smith, A., Watkiss, P., Tewddle, G., McKinnon, A., Browne, M., Hunt, A., Treleven, C., Nash, C. and Cross, S. (2005) *The Validity of Food Miles as an Indicator of Sustainable Development* London: DEFRA.
Soil Association (2002) *Organic Food and Farming Report 2002* Bristol: Soil Association.
Southerton, D., Chappells, H. and Van Vliet, V. (2004) *Sustainable Consumption: The Implications of Changing Infrastructures of Provision* Aldershot: Edward Elgar.
Thompson, C. J. and Arsel, Z. (2004) 'The starbucks brandscape and consumers' anticorporate experiences of globalization' *Journal of Consumer Research* 31, pp. 631–642.
UNCED (1992) *Agenda 21: The United Nations Program of Action from Rio* New York: U.N. Publications.
Van Vliet, B., Chappells, H. and Shove, E. (2005) *Infrastructures of Consumption: Environmental Innovation in the Utility Industries* London: Earthscan.
Wackernagel, M. and Rees, W. (1996) *Our Ecological Footprint: Reducing Human Impact on the Earth* Philadelphia: New Society Publishers.
Ward, B. and Lewis, J. (2002) *The Money Trail* London: New Economics Foundation.
Whatmore, S. and Thorne, L. (1997) 'Nourishing networks: Alternative geographies of food', in D. Goodman and M. Watts (eds) *Postindustrial Natures: Culture, Economy and Consumption of Food* London: Routledge, pp. 287–304.
Winter, M. (2003) 'Embeddedness, the new food economy and defensive localism' *Journal of Rural Studies* 19(1), pp. 23–32.

Part IV
Conclusion

Conclusion

Martin Ryle, Kate Soper and Lyn Thomas

Several of the contributors have responded to our invitation, as the book goes to press, to add brief concluding reflections to their chapters, commenting on how these are related to the project of the volume as a whole and to the political moment in which it is appearing. Here, the editors reflect collectively on the questions the book raises – not in the desire or expectation of drawing up a final balance-sheet, but to emphasise the scope of the discussion which we hope it will help to promote. Kate Soper in her Introduction has charted the complex relationship of new counter-consumerist and alternative hedonist values and projects to some earlier debates about consumption in the 'West'; in the light of that history, we offer a last word about the present and the future political implications of the book. For while many of the chapters focus on well-defined empirical and cultural aspects of contemporary consumption and on particular theoretical approaches to it, taken together they raise very general concerns about the role of the consumer as the agent who can sustain, modify, trouble or perhaps ultimately destabilise the workings of the contemporary globalised economy.

Simplifying somewhat, we can suggest that three distinct political modes or horizons are represented. There is discussion, first of all, of a range of campaigns and initiatives ('Fair Trade' and 'ethical consumption') that seek to alter the consumer-producer relationship, through better-informed and more discriminating shopping. Sam Binkley, Jo Littler and Roberta Sassatelli discuss some of the questions that arise, and suggest some of the institutional and informational basis needed if these kinds of movement are to make the difference their adherents desire. Already, international trade in certain goods (particularly agricultural produce and clothes) is increasingly under pressure to adapt itself to the preferences and demands of relatively well-to-do 'new consumers'. Something more than a niche market may be developing: for example, posters in the streets of Dublin were proclaiming the whole city's 'Fair Trade' credentials during March 2008. However, the level and content of consumption is not itself at issue: these kinds of campaign seek to ameliorate the terms of international trade, but they do not aim to reduce it. Some commentators have even argued that this is a new phase of consumerism, too readily appeased by corporate PR and 'greenwash'.

Gill Seyfang's chapter offers a detailed presentation and analysis of a second mode of politicised consumption, answerable mainly to environmental rather than ethical-humanitarian concerns, which seeks to develop and sustain local trade in food as a way of promoting the local and regional economy and radically reducing the 'food miles' entailed by its globalised alternative. Seyfang's account stresses that this kind of principled localism can run parallel to associated Fair Trade initiatives, comments on its relationship to the organic movement and highlights its explicitly cultural-political motivations. Here, a significant shift in the pattern and content of trade is envisaged, not least because while 'food miles' represent from one point of view a waste of energy and labour, from another they represent a source of profit and employment for the companies and individuals who currently transport food around the world. Seyfang herself suggests that the initiatives she is charting point logically towards the eventual replacement of the existing supermarket-based international food supply chain by new institutions and infrastructures. She does not explicitly raise the further question of how far such an expansion of localised trading arrangements, at the expense of the major corporations that presently dominate food retailing, might in the long run adversely affect the profitability of an important sector of the mainstream economy.

Anxieties about adverse economic effects of alternative consumption can be recast, of course, in positive or utopian terms: as questions about what a different and better economic system might look like. Can we imagine, and might we one day institute, another economy, which would thrive on a less intensive use of resources and labour, and would no longer need continuous overall growth (which is presently the basis not just of corporate and company profits, but of the taxation that supports expanding public services)? That large question is undoubtedly raised by this book, inasmuch as its predominant perspective, especially in the chapters of cultural criticism and general evaluative reflection, focuses on not consuming, or consuming less, as well as on consuming differently. In this third mode, the 'politics of consumption' is not just about ethically and environmentally motivated shopping, it is about sometimes choosing not to shop. It is a central contention of this book that this is not a purely negative decision, but one expression of a desire for another mode of life, whose dimensions and representations many of the chapters trace. Above all, we want to raise the question of whether a society of lower consumption might not also be a more pleasurable society.

Ambivalence about consumption and principled abstention from it tend – implicitly, and sometimes by explicit intention – to run counter to established economic priorities and imperatives; in that sense, they gesture towards the inauguration of an alternative order in the realm of production. However, this does not mean that we can readily designate such an alternative economic system, or that the political goal of finding one is generally pursued even within counter-consumerist circles. When the question of consumerism and the environment was raised by the 'new social movements' of the early 1980s, it was generally framed within a broadly socialist critique of capitalism. Since then, the adoption of neo-liberal economic policies by most nominally left-of-centre parties and governments has made appeals to socialist traditions appear quixotic. The question of a non-expansive mode of production may still haunt the historical horizon, with many even of the most orthodox representatives of neo-liberalism acknowledging the pressures of resource depletion and the perils of our ever-larger carbon footprint (even if they scrupulously avoid any suggestion that the pursuit of endless growth might be contributing to the emergency). But it is easy to feel that it has become purely abstract: there currently seem to be no credible answers as to what institutional forms a non- or post-capitalist economy might take, or what actors and agencies – political parties, social movements, labour organisations, enlightened governmental and business elites – might pursue it.

One should, however, acknowledge the work of those whose investigations of sustainability may open up these larger, systemic questions: the Introduction mentions (in Britain) the work of the Sustainable Development Commission, and the ESCRC RESOLVE research programme at the University of Surrey. In the future, such initiatives are likely to become more widespread and influential. Of course, any transformation of the global market, even if we envisage it as a process of adjustment over time, would be most contentious, and could only happen if fundamental economic questions returned to the centre of the political stage in the world's richest democracies. Readers of this book, and indeed the contributors to it, will differ as to how far its critical discussion of consumption implies the eventual need for such a political and economic transformation. For our part, we note Roberta Sassatelli's thoughtful concluding reflections, which underline both the potential and the necessary limits of politically motivated consumption, and suggest its place in the wider democratic public sphere. 'Politics and consumption', she writes, 'can act in synergy in the transformation of the market.' Those who agree (as we do) with Sassatelli may conclude that the interventions collected here point beyond the consumer as individual shopper, towards the citizen as one whose consumption is part of a collective critical and political practice.

Index

Abbott, J., 83
ABC listing, 62
Abram, D., 134–7, 139, 140, 151
Accidental, The, 51, 53–5
Achbar, M., 83
action, 122
active citizenship, 192
activism, civic media, 83
Adams, J., 157
Adbusters, 93
Addiopizzo!, 33
Adorno, T., 9
advertising, commercial, 30
  analyses of, 70–1
  lifestyle magazines, 60–1
  postmodern, 32
  pro-consumerism, 27, 29, 33
  US, 31
  vulnerability of, 31–2
Advertising Women of New York, 31
aesthetic manipulation, 105
agency
  political, 57
  of working class, 10
Akass, K., 184
Alan, F., 12
Alibhai-Brown, Y., 80
alienation, modern, 134
Altermarket, 33–8
alternative hedonism, 3–6, 66, 107–8
altruism, moral, 5
American Apparel, 37, 180
'American Way of Life', 30
Andreas, F., 174
Anthony, G., 94
anti-consumerism, 28, 30, 93, 100, 173
  gendered, 176–7, 184
  strategies, 95
apocalyptic views, 33, 38
Appleby, J. O., 27
Arendt, H., 114, 122
aristocracy, symbolic, 77
Aristotle, 114, 115
Armstrong, F., 83
Arnould, E. J., 34, 39
Arsel, Z., 106, 195
art of living, 1, 113, 114–16
Athanasiou, T., 82
auditing, ethical consumer choices, 109

Bad News, 78
Bahro, R., 9
Barker, A., 79
Barnett, C., 11, 37
Barnett, S., 78
Barry, J., 125, 126
Barton, H., 192
Baudrillard, J., 25
Baudrillard, J., 25, 26, 36
Bauman, Z., 94
Beck, U., 8, 10, 96, 118–19
Beder, S., 82
Beer, G., 47
'being'/'having', 121
Bell, D., 79, 104, 171
Bello, W., 81
Benetton, 32
Benkler, Y., 83
Bennett, L., 97, 98
Benson, A. L., 158
Benton, T., 9
Berg, M., 75
Berger, J., 77
Beyond Boredom and Anxiety, 164
Binkley, S., 15, 93, 100, 209
biosemiotics, 141
Black Friday, 93
Blair, T., 80
Blanchard, S., 14, 59, 74, 78, 82, 83
Blumler, J., 83
Blyth, M., 77, 78
bodily hexis, 101

body
  the bohemian, 106
  as object of reflexive
    self-awareness, 99
  pornographic, 54
Body Shop, 31
bohemian habitus, 104, 106
bohemianism, 103–7
bohemians, urban, 103
Boltanski, L., 27, 35
Bond, P., 80, 81
Boorstin, D., 76
borders, in time/space, 46–9
Bordo, S., 29
Bourdieu, P., 12, 76, 95, 99, 101, 102,
  104
Bowlby, R., 30, 174
boycotts, consumer, 176, 177
Boyle, D., 191
Brave New World, 54
Breward, C., 184
Brewer, J., 75
Bronte, C., 47
Brown, K. W., 2, 181
Brown, W., 182
Browning, R., 160, 161
Brutus, D., 80
'bulimic personality type', 29
Bunting, M., 4, 6
bureaucratisation, 28
Burgess, J., 190
Burton, R., 162
businesses, environmentally
  responsible, 71
Butler, J., 178, 180
Buy Nothing Day, 93
Byatt, A. S., 47

Cairncross, F., 167
Callon, M., 38
Cammack, P., 81
Campbell, C., 16, 117–18, 142, 145,
  146
capitalism, 10, 44, 45, 103, 118
Carlson, L., 37
Carpenter, H., 77
Carrabine, E., 158
Carrier, J., 38
cars
  as commodity, 157–8
  impact of, 157
Carson, R., 75
Carter, E., 181
Castells, M., 29, 83
Ceccarini, L., 33
Center for the New American
  Dream, 6
change, social/ socio-cultural, 63, 94
Chatwin, B., 162
Chessel, M-E., 33
citizen-consumer, 11
citizenship, 39
  active, 192
  ecological, 126–7
civic participation, 36
Clarke, J., 27
Clarke, N., 27
class, creative, 103–4
climate change, 3
  catastrophic, 147–9
Cloke, P., 37
cloning, 52
Cobb, J., 191
Cochoy, F., 33
Cockett, R., 78
Cohen, E., 25, 27, 34
Cohen, L., 177
Coleridge, S. T., 143, 150
collective action, 192–3, 200
collective goods, 35
Collett, P., 158
Collini, S., 77
Coming up for Air, 48
commercial advertising, 30
  analyses of, 70–1
  lifestyle magazines, 60–1
  postmodern, 32
  pro-consumerism, 27, 29, 33
  US, 31
  vulnerability of, 31–2
Commission for Africa, 80
commodification, 100
  Marxist/Frankfurt School views, 9
  and othering, 32
commodity chains, 32
community action, 202
community-building, 192, 198–9

compulsory consumption, 184
Conca, K., 94
conspicuous compassion, 81
consumer choice and voting, 36
consumer habitus, 99–102, 103
consumerism
  apocalyptic views, 38
  attitudes to, 4
  as civilising force, 27
  ecological, 126–7
  and ethical norms, 50
  and gender, 174–6, 182, 184
  and happiness/well-being, 3
  political, 94, 102, 107, 108, 109
  problems caused by, 62
  romantic, 141–6
  and simple life, 71
  and spiritual void, 60
consumer–producer relationship, 209
consumers
  ethically oriented, 37
  as global actors, 36
  images of, 25
  political-economic role, 27
  western middle-class, 31
consumer society, 74
consumer sovereignty, 27
Consumer's Republic, A, 177
consumer whore, 172
consumption, 27
  apocalyptic/apologetic views of, 29, 33
  and democracy, 38
  erotics of, 108
  ethical, 11
  Euro-American mode, 3
  and gender, 173, 184
  greening of, 105
  institutional, 190
  and movement, 158
  politics of, 9, 33
  as private matter, 25, 57, 118
  promotion of, 35
  and sociological theories, 94
  sustainable, 188, 189–91
  Western first-world, 55
contradictory moments, 55
cooperatives, 34, 200
Corner, J., 83
counter-consumerism, 4, 7–12, 59
counter-culture, 30
Country and the City, The, 44, 46
Country Living, 60–2, 64–8, 70
Craig, L., 70
Crisell, A., 77
Cromwell, D., 80
Crouch, C., 83
Csikszentmihalyi, M., 164
Csordas, T. J., 102
cultural mediation, and economic development, 103
cultural shifts, 72
Cultural Studies, 183–4
culture
  material, 28
  mediated, 75–6
Cultures of Consumption Programme, 7
culturization, of economy, 103

Dalton, R., 94
Daly, H., 191
Darlow, M., 78, 79, 80
Daunton, M., 11, 25
Davis, A., 82
Deakin, R., 130, 131, 132, 133
Debord, G., 76
de Geus, M., 15, 113, 115, 119
de Graaf, J., 6
De Grazia, S., 165–6
de Grazia, V., 174–5
deliberation, 123
democracy, 38, 52
desire, 3, 117–18, 142
desire-acquisition-use-disillusionment-renewed desire, 117
development, self, 52
De Zengotita, T., 76
Dickens, C., 47
Diener, E., 1
Dinan, W., 82
distribution channels, new, 194
distribution networks, alternative, 36
'ditch vision', 133, 145
Diversity of Life, The, 75
Dobson, A., 118, 120, 126, 192
documentaries, 83

Dohmen, J., 114
Don, M., 59
Doubleday, R., 40
Douglas, M., 29
Douthwaite, R., 189, 191, 192
Dovey, J., 79
downshifting, 6, 35
Dubouisson-Quellier, S., 34
Du Gay, P., 103
DuPuis, M., 189, 196
Dwyer, K., 172
Dyke, G., 80
Dystopia, holidays, 67

East Anglia Food Links (EAFL), 200
ecochic hotels, 64, 71
eco-clothing, 179–80
ecological
  citizenship, 126–7
  consumerism, 126–7
  crisis, 147–9
  hedonism, 123–4, 131
  space, 192
  virtues, 125–6
economic development, 103
economic imperatives, structural, 56
Economies of Signs and Space, 103
ecophenomenology
  and pleasure, 134–7
  and revelation, 137–41
eco-reality, 60
eco-tourism, 49–51, 131
education, ethical consumer choices, 109
Edwards, D., 80
Ekins, P., 191
Elgin, D., 6
élite neoliberalism, 82
élite networks/agendas, 81
Ellis, J., 79
embodiment, 99–102
Emecheta, B., 47
Entwistle, J., 95
environmental
  catastrophe, 3
  destruction, 43
environmentalism, 74–5, 131, 134
Eostre Organics, 196–7
eroticism, commodified, 55
erotics of consumption, 108
escape, 66–71
ESRC RESOLVE programme, 211
ethical
  consumerism, 107
  consumption, 11
  shopping, 37
Ethics, 115
Euro-American mode, of consumption, 3
European Social Survey, 34
Evans, D., 5
Evans, P., 84
evolutionary psychology, 150
Ewen, S., 30
exemptionalism, 74–5, 81, 82
exploitation, social, 4

'fair Earthshare', 125
Fair Trade
  Italy, 36, 38
  problems in, 37
  sales of, 33–4
fantasies, mobilisation of, 71
'farmers' markets, 194
Fearnley-Whittingstall, H., 59
Featherstone, M., 105
feel-good emotions, 81
femininity
  and consumerism, 17, 181
  labour of, 70
Fenkl, H. I., 172
Filkin, G., 191
Finch, J., 64
Fine, B., 188, 190
Fish, R., 61, 65
Fisher, F. J., 75
Florida, R., 95, 103–4
flow activities, 164, 169
Flower Power, 9
food
  miles, 194, 197–8, 210
  networks, embedded, 195
  organic/local, 189
footprint, ecological, 124–5, 192, 198
Forno, F., 33
fossil fuels, 84
Franche-Comté, 51

Frankfurt School, and commodification, 9
Frank, T., 30, 31
Freedom, of choice/action, 118
Frey, B., 1
Friedland, L., 37
Friedman, M., 34
Fromm, E., 114, 120, 135
frugality, pleasures of, 35
Fuller, G., 77
Furedi, F., 77
Furlough, E., 34, 174, 175
Future, impact on past, 53

G8 process/summit, 80
Gabriel, Y., 177
garment workers, 37
Garrard, G., 43, 47
Gelpke, B., 83
gender, and consumerism, 17
gendered overconsumption, 183
geneologies, alternative, 176–8
Gernsheim, E., 36, 96
Gibney, A., 83
Giddens, A., 10, 94, 95, 97, 98, 99, 100
Gifford, T., 43, 44, 51, 54
Gill, S., 17, 188, 210
Ginsborg, P., 11
Gissing, G., 47
Glasgow University Media Group, 78
global corporations, 62
globalisation, 195
global warming, 50
Goddard, P., 77
Goeddertz, T., 32
'Going Green', 63
'Golden Ages', 43, 45
Goodin, R. E., 82
good life, the, 49, 113, 114–15
Goodman, D., 34, 189, 196
Goodwin, P., 78
Gore, A., 149
Gorz, A., 9
Gough-Yates, A., 61
governance
  institutions, 188
  mediated, 83
Granovetter, M., 77
gratification, in objects, 28
Greene, G., 84
'green' practices, as accessories, 71
Greenwald, R., 83
greenwashing, 82, 109, 210
Griffiths, J., 134, 136
Grosz, E., 178, 179, 181
growth, economic, destructive consequences, 55
Guggenheim, D., 84
Guthman, J., 199

habitus, 95
  bohemian, 102–7
  consumer, 99–102, 103
Halkier, B., 97
happiness, 1, 3, 29, 120
Happy Planet index, 1–2
Haraway, D., 173
Hardy, T., 46, 47
Hari, J., 80
Harrison, B., 77
Harrison, R., 11
Harvey, D., 78, 95, 97, 103, 140
Harvie, D., 81
Haug, W., 9
'having'/'being', 121
Hay, C., 77, 83
Hearst Corporation, 62
Heath, J., 106
hedonism
  alternative, 3–6, 66, 107–8
  contemporary/modern, 117, 142
  ecological, 123–4, 131
  sustainable, 119–21, 124, 127
Hegland, J., 133
Henderson, H., 191
Hennesey, R., 183
Hewison, R., 44
Heyck, T. W., 77
Hilton, M., 11, 25, 28, 177
Hinrichs, C. C., 195
Hirsch, F., 29
Hirschman, A. O., 27, 28
historical choices, 53
Hobbes, T., 114, 115, 116, 118
Hodgkinson, T., 4
Hodkinson, S., 80
Holdsworth, M., 190
holidays, problems with, 66–7

Holley, D., 166
Holloway, L., 189, 194
Hollows, J., 79
Holst, C., 6
Holton, R., 102
Holzer, B., 94
home
  concepts of, 47
  spaces of, 68, 70
Homes and Gardens, 60–2, 64, 68–9
Honore, C., 6
Hooker, J., 132
hope, 167, 169
Hopkins, R., 84
Horowitz, D., 28
Houellebecq, M., 13, 44, 49, 50, 55, 56
Hubbard, G., 81
Hudson, I., 38
Hudson, M., 38
Human Condition, The, 122
Human Development Index (HDI), 5
humour, 137
Huxley, A., 52, 54
Huyssen, A., 174
hyperconsumerism, as ressentiment, 180–4
hypermobility, 168

Ifoam, 34
image society, 76, 77, 82
'Immortality Ode', 48
impoverishment, spiritual, 28
Inconvenient Truth, An, 149
Index of Sustainable Economic Welfare (ISEW), 5
individualisation, 96–7
individualism, possessive, 116
inequalities, 17, 183
infrastructure, social, 189
Ingold, T., 137–41
institutions, new, 201
institutions, new socio-economic, 193
intellectual field, TV as, 76
intellectuals, 77
Intergovernmental Panel on Climate Change, 147
Iraq, 80
Isaacs, J., 78, 80
Isherwood, B., 29
Ishiguro, K., 13, 44, 51, 52, 56
Italy, 32, 33, 34–5, 36
It's not easy being green, 60

Jackson, T., 2, 5, 188, 191, 192, 193
Jacobs, J., 191
Jain, J., 166
Jameson, F., 9
Jannson, A., 76
Jasanoff, S., 189
Jones, A., 189, 194
Jones, J., 183
Jones, P., 157
Jones, P. T., 124–5

Kampfner, J., 80
Karliner, J., 82
Kasser, T., 2, 5, 6, 120, 121, 122
Kauppi, N., 101
Kavanagh, D., 83
Kerridge, R., 16, 44, 130
Kier, J. M., 33–4
Klein, N., 94
Kluger, A. N., 166
Kneafsey, M., 189, 194
Knight, G., 97
Koslowsky, M., 166
Kraidy, M. M., 32
Kristen, L., 36
Kunstler, J. H., 84

Labelling, product, 109, 125
Labour governments, and socialism, 45–6
labour/work/action, 122
Lamine, C., 34
Lanchester, J., 147–8
Landes, J., 183
landscape, 138
  see also nature
Lane, R., 2
Lasch, C., 28
Lash, S., 103
Lasn, K., 93, 94
La société de consommation, 25
Lawrence, D. H., 46–7

Layard, R., 2, 192
left-wing, and consumers/agency, 10
Leiss, W., 9
Lemaire, T., 123–4
Leonini, L., 35, 36, 37, 38
Leopold, E., 188, 190
Lerner, L., 43
Levett, R., 4, 5, 188, 190
Leviathan, 116
Levi-Faur, D., 81
Levi, M., 38
Lewis, J., 195
Lewis, T., 111
Leys, C., 78
Leyshon, A., 190
life politics, 10, 36, 95–9
life space, 165–6
lifestyle magazines
  advertising/advertorial content, 60–1
  escaping from the escape, 66–71
  'greening' of, 62–6
  readership, 61–2
lifestyles, 69, 120
Linton, A., 38
literary fiction, 53, 56
literature, and understanding, 44
Littler, J., 17, 37, 100
Live, 8, 80–2
Livingetc, 60–2, 64, 68
Lloyd, R., 103
Loach, K., 83
localisation, 191–2, 197–8
localism, 195
  reflexive, 196
Locke, J., 114, 115–16
Lockie, S., 36
Lodziak, C., 9
Long Emergency, 84
Longhurst, B., 158
Louw, E., 76
Lovelock, J., 147, 149, 150, 152
'Love your body' campaign, 31
Lukács, G., 28
Lury, C., 71
Lutz, M., 191
Lynas, M., 147
Lyons, G., 166
Lyons, J., 171
McCormack, R., 84
McCormick, J., 75
McGahern, J., 47–9, 51
Machin, D., 76
McIlroy, J., 47
McKibben, B., 132, 149
McLaren, C., 180
McLuhan, M., 76
McRobbie, A., 103, 180, 183, 184
Mafia, and advertising, 32
Mair, P., 83
Make Poverty History, 80
Maniates, M., 188, 190, 192
manipulative populism, 83
Marcuse, H., 9
market, alternative views, 34
marketing, 27, 30, 109
market segmentation, 61
market society, 9, 25
Marks, N., 2
Marsden, T., 194
Marshall, G., 83
Marsh, P., 158
Martello, M., 189
Marxism, and commodification, 9, 28
Masculinity, and consumerism, 184
Maslow, A., 2, 164, 165
materialism, 121–2, 142
Matless, D., 131
Mayo, E., 183, 184
Medawar, P., 163
media
  counter-consumerist, 59
  and politics, 82
mediated governance, 83
mediatisation, 75–6
Memoir, 48, 49
memorialisation, 123
Merchant, C., 84
Merleau-Ponty, M., 135–6, 137, 139, 161
Mertes, T., 84
Met Open Zinnen; Natuur, Landschap, Aarde, 123
Meyer, T., 83
Micheletti, Michele, 97–8, 109, 176, 177
Milburn, K., 81

Miller, D., 6, 11, 29, 81, 82
Minney, S., 179
'modern', 52
Modernisation, industrial, 96
modernity, 27
modernization, reflexive, 9
Mokhtarian, P. L., 157
Monastery, The, 60
Monbiot, G., 63, 82, 149
Moor, L., 37, 180
moralistic approach, 118–19, 127
Morgan, F., 84
Morley, D., 78
Morris, M., 32
mortality/death, 52
Moseley, R., 79
motivation
  hedonistic, 5
  to sustainability, 3
movement, 167
  and pleasure, 158
  primacy of, 159–60
Murdoch, J., 189, 194

narcissism, 28
National Consumer Council, 184
National Consumer League, 176
National Magazine Company, 62
National Readership Survey, 61, 62
nature, 130–4, 144
  see also landscape; rural life
Nature-Deficit Disorder, 133
Nava, M., 174, 176
Nehamas, A., 114
Nelson, F., 65
Nelson, M. R., 35
neoliberalism, and consumers/agency, 10
neoliberalism, élite, 82
Never Let Me Go, 51–3, 56
New Economics, 17, 188
  and sustainable consumption, 191–3
New Economics Foundation, 191
  Happy Planet index, 1
New Right, 78, 79
Nietzsche, F., 181
Nixon, S., 31
Norberg-Hodge, H., 189, 195
Norfolk, 51
Nostalgia, feminine aesthetics, 70
No Waste Like Home, 60
Nussbaum, M., 35, 114

Oborne, P., 80, 83
occupational structures, 103
O'Connor, J., 9
Office of Communications (Ofcom), 79
Oliver, J., 59, 200
Oliver, M., 77
O'Neill, J., 2
Open Senses; Nature, Landscape, the Earth, With, 123
organic/local food, 189, 193–6
organic production, 34, 36
O'Riordan, T., 192
Orwell, G., 48
othering, and commodification, 32

Paddock, C., 36
Palmer, G., 79
Parks, L., 184
Pascal, B., 115
pastoral, 43–4, 56
  exotic, 50
  power, 54
  and return to city, 54
patriotic shopping, 2
Patterson, C., 47
Paxman, J., 80
Payne, A., 80, 81
Peak Experience, 164, 165
Pearce, F., 147
Pels, D., 83
Pemberton, H., 77
perception, 135–6
Perception of the Environment, The, 137
personalisation, 100
phenomenology, 140
Pierce, L. B., 6
places, marginal, 50
Platform, 49–51, 55

pleasures
  ecophenomenological, 134–7
  forms of, 43, 124
  green, 131
  human, 4
  and lifstyle magazines, 66
  of movement, 160–5
  of travel, 165–6, 169
Plumb, J. H., 75
Pocock, J. G. A., 27
policy approaches, 190
political agency, 57
political consumerism, 94, 102, 107, 109, 173
  gendered, 177–8
  as social movement, 108
political consumption, 178–80
  and gender, 175
political investment, consumption, 33
politics
  of consumption, 9, 212
  of food production, 59
  gesture, 80
  global, 81
  life/life-style, 10, 36, 95–9
  location of, 96
  and media, 82
  post democracy, 83
poor, the, 168
Popper, K., 119
popwash, 82
Porritt, J., 190
Porteous, T., 81
Porter, R., 75
positivism, in sociology, 8
possessive individualism, 116
post democracy politics, 83
postmodernism
  and advertising, 32
  and consumer culture, 11–12
  and consumers/agency, 10
poverty, 168
power, pastoral, 54
Preston, A., 80
Pretty, J., 189, 194
private/public realms, 122
pro-consumerism, 25, 27, 30
production, non-expansive, 211
productive efficiency, 46
prosperity, redefinition of, 5
psychology, evolutionary, 150
public relations, corporate, 81–2
public sector
  catering, 200
  and consumption, 190
public service ideals, television, 79
Purdy, D., 2, 4

quality of life, 35, 66
  measuring, 36
Quality of Life index, 5
Quarmby, K., 80

Radner, H., 174
rationalisation, 28
'real beauty', 31
Redclift, M., 9
Redefining Prosperity, 2
red–green formation/literature, 9, 10
redistribution/interdependency, 35
Reed, M., 193, 194
Rees, W., 124–5
reflexive modernity, 96, 97, 98
reflexive modernization, 9
reflexivity, 99–102
regimes of justification, 35
Reich, M., 166
Renting, H., 189, 195
research
  Cultures of Consumption Programme, 7
  Italy/consumer sovereignty, 34–5
RESOLVE programme, 211
Responsible Consumerism, 158
ressentiment, 180–4, 185
rest, 166–7
retail therapy, 158
  see also shopping
retrospective radicalism, 44–6
Revenge of Gaia, The, 152
Ricketts, H. J., 194, 197
Risk Society, 118
Robertson, J., 191
Rojek, C., 94
Rolt, L. T. C., 161
Romantic Ethic and the Spirit of Modern Consumerism, The, 117, 142

romanticism, 132, 137, 139, 141–6, 152
Ross, A., 130
Rousseau, J-J., 116
Rowell, A., 82
Roy, A., 172
rural
  life, idealisations of, 44
  pleasures, lost, 48–9
  simplicity, 54
  see also landscape; nature
rural-intellectual radicalism, 45
ruralism, conservative, 45
Ryle, M., 9, 10, 13, 46, 49

Safe Lanes scheme, 65
Saint-Exupéry, A., 164
Salamon, I., 157
Sam, B., 15, 93, 209
Sanne, C., 188, 190
Sassatelli, R., 11, 13, 28, 30, 33, 35
Scammell, M., 36
Schatzki, T., 12
Schivelbusch, W., 165
Schor, J., 4, 6, 28
Schudson, M., 36, 37
Schumacher, E. F., 191
Seamon, D., 167
Search of Nature, In, 74
Searle, G., 28
Second Treatise, 116
self-actualisation, 98, 100
self-awareness, reflexive, 107
self-realisation, 4, 100
Seligman, M., 1
Sen, A., 35, 114
Sen, A. K., 29
sensation, physical, 160–5
Setshedi, V., 80
sex, as commodified eroticism, 55
Sex of Things: Gender and Consumption in Historical Perspective, The, 175
Seyfang, G., 188, 190, 193
Shah, D. V., 37
Shah, H., 4
shopping, 2, 11, 37, 158
Shopping Generation, 184
Simms, A., 192
simpler life, 71
Sklair, L., 81
'slave morality', 181
Slow City network, 6
Slow Food network, 6
Smith, A., 53, 54, 56, 193, 194, 195
Snoddy, R., 81
social
  change, 94
  exclusion, 192
  exploitation, 4
socialism, 45–6
social movements, new, 84, 98
social practice, theory of, 101
societies, market, 25
society
  consumer, 74
  image, 76, 77, 82
  market, 9
  work-driven, 6
socio-economic institutions, 193
sociology, positivism in, 8
Solomon, J., 16–17, 157
Soper, K., 1, 10, 11, 35, 36, 66, 131
Southerton, D., 189, 190
spaces
  concepts of, 51
  of home, 68, 70
  personal, 169
  of political action, 122
Space, Time and Perversion, 179
Spell of the Sensuous, The, 135, 136
spiritual impoverishment, 28
spiritual naturalism, 123–4
standardisation, 28
Starbucks, 171
Stay Free!, 180
Stolle, D., 94
Stolle, S., 37
Strasser, S., 25
stress, 67, 121
Strikwerda, C., 34
Stutzer, A., 2
subpolitics, 95–9, 97
Suistainable Devlopment Commission, 2
supermarkets, 201, 210
sustainable consumption, 188

Sustainable Development Commission, 190, 211
  Report 2003, 5
sustainable hedonism, 119–21, 124, 127
Swadeshi movement, 176
sweatshops, 37
swimming, 130–1, 133, 134
Switzer, J. V., 82
symbolic aristocracy, 77
systems of provision, 188, 189
  alternative, 190–1, 193

Talen, B., 171, 172
'tasks', 138, 139
Teenagers, and consumerism, 183
telegentsia, 74, 76–8
television, 74, 76, 77, 78–80
Terra Libera!, 33
Tesco (magazine), 60–2, 63, 66
Tesco (organisation), 63
That They May Face the Rising Sun, 49
Thévenot, L., 27, 35
Thomas, L., 4, 7, 14, 59, 209
Thompson, C., 106, 195
Thorne, L., 195
time
  affluence, 6
  spatialisation of, 51
Time Warner Corporation, 62
Tosi, S., 33
Touraine, A., 95
tourism, 49–51, 131
transnational capitalist class, 81
transport, 157
travel
  meaning of, 159
  pleasures of, 165–6, 168
Trentmann, F., 11, 25, 34
Tunstall, J., 76
turbo consumerism, 84

United Nations Conference on Environment and Development, 75
urban–rural relations, 44
Urry, J., 103
US consumers, 30

Van Vliet, B., 189
Veblen, T., 28
virtues, ecological, 125–6
virtuous shopping, 11
vita activa, 122
Voluntary Simplicity movement, 6
Voting, and consumer choice, 36

Wackernagel, M., 84, 124, 125, 192
Wacquant, L., 102
Walsh, M., 94
Waltz, M., 83
Ward, B., 195
Warde, A., 12
War on Terror, 80
Waterlog, 130
wealth
  divisions of, 1
  and happiness/well-being, 121
websites, progressive, 84
well-being, 1, 3, 29, 35
Wells, H. G., 168
West, P., 81
Whatmore, S., 195
Wheeler, W., 137, 140, 141, 146
Where Have All The Intellectuals Gone?, 77
Whitehouse, M. S., 159, 163
White Label campaign, 176
Whiteside, K. H., 123
Whole Creature, The, 140
Williams, C., 183
Williams, C. C., 36
Williams, C. L., 183
Williams, P. D., 80–1
Williams, R., 10, 13, 43, 44, 46, 47, 52
Williamson, J., 71
Willis, P., 30
Wilson, C., 164
Wilson, E., 104
Wilson, E. O., 74–5, 82, 141, 143
Winship, J., 181, 183
Womack, A., 104
Women and the Public Sphere in the Age of the French Revolution, 183
Woodgate, G., 10

Woodward, J., 78
Wordsworth, W., 48, 141–2, 143
work, 122
workforce, 103
working class, prosperity of, 10
World Trade Organisation, 34
Wrigley, E. A., 84
Wrigley, T., 84

Zinkhan, M. G., 37
Žižek, S., 148–9
Zukin, S., 103